Realizing Strategy Through Projects

The Executive's Guide

Best Practices and Advances in Program Management Series

Series Editor
Ginger Levin

RECENTLY PUBLISHED TITLES

Realizing Strategy through Projects: The Executive's Guide
Carl Marnewick

PMI-PBA® Exam Practice Test and Study Guide
Brian Williamson

Earned Benefit Program Management: Aligning, Realizing, and Sustaining Strategy
Crispin Piney

The Entrepreneurial Project Manager
Chris Cook

Leading and Motivating Global Teams: Integrating Offshore Centers and the Head Office
Vimal Kumar Khanna

Project and Program Turnaround
Thomas Pavelko

Project Portfolio Management in Theory and Practice: Thirty Case Studies from around the World
Jamal Moustafaev

Project Management in Extreme Situations: Lessons from Polar Expeditions, Military and Rescue Operations, and Wilderness Exploration
Monique Aubry and Pascal Lievre

Benefits Realization Management: Strategic Value from Portfolios, Programs, and Projects
Carlos Eduardo Martins Serra

IT Project Management: A Geek's Guide to Leadership
Byron A. Love

Situational Project Management: The Dynamics of Success and Failure
Oliver F. Lehmann

Ethics and Governance in Project Management: Small Sins Allowed and the Line of Impunity
Eduardo Victor Lopez and Alicia Medina

Realizing Strategy Through Projects

The Executive's Guide

Carl Marnewick

CRC Press
Taylor & Francis Group
Boca Raton London New York

CRC Press is an imprint of the
Taylor & Francis Group, an **informa** business
AN AUERBACH BOOK

CRC Press
Taylor & Francis Group
6000 Broken Sound Parkway NW, Suite 300
Boca Raton, FL 33487-2742

Printed on acid-free paper

International Standard Book Number-13: 978-1-138-19610-0 (Hardback)

Visit the Taylor & Francis Web site at
http://www.taylorandfrancis.com

and the CRC Press Web site at
http://www.crcpress.com

Dedication

To Annlizé and Lyoné

Contents

Preface

This book is written for current and prospective organisational leaders and executives who want to understand the role and importance of project management.

The book focuses on the value and benefits of project management to the organisation. The ultimate goal is to provide organisational leaders with a view as to how project management can deliver organisational strategies. Project management as a discipline does not exist in isolation. Project management should be seen within the perspective of portfolio and programme management as well as from the perspective of the role that it plays within the larger organisation. Specific aspects within the project management discipline are focused on in terms of how each of them should be managed from a business perspective rather than from a project management perspective.

Various books either cover the project management discipline in depth or discuss specific aspects of project management. This book is different in that it provides a holistic view of what constitutes project management from an executive perspective. This perspective can be used to purposefully position project management within the organisation as a strategy enabler.

This book covers the entire spectrum of project management from a management and leadership perspective. Its focus is not necessarily on what needs to be done, from a project management perspective, but what organisations and senior executives can do to make the facilitation of projects easier. The first aspect that is addressed is the value of project management. The focus is on value-driven project management and its rationale. The second aspect focuses on project management as a strategic enabler. Of the three disciplines—portfolio, programme, and project management—project management is at the lowest level. Portfolio and programme management are elaborated on in detail. This is to enable the reader to understand the symbiotic relationship between portfolio, programme, and project management. These three concepts are explained from the view of the organisational leaders. The role of the project management office (PMO) as an organisational institution that can facilitate the successful delivery of projects, programmes, and portfolios is also elaborated on.

The concepts of benefits realisation management and sustainability from a project management perspective are also covered. At the end of the day, projects deliver a service or product. The focus of benefits realisation management is to determine how the associated benefits of a service or product can be harvested and sustained for the benefit of the organisation. Sustainability is currently a very relevant topic, and how sustainability can be incorporated into project management is addressed. This is done through the perspective of people, planet, or profit, or the so-called triple bottom line.

Governance, and especially project governance, plays an important role in today's organisations. This is evident from all the corporate scandals that have made the headlines over the last couple of years. This book focuses on governance within the project management domain. Governmentality, project governance, and governance of projects are discussed as well. This book concludes with the role of the executive sponsor and how the executive sponsor contributes to the success of projects, project management, and the organisation at large.

Acknowledgements

I wish to extend my gratitude to Dr. Ginger Levin, who convinced me of the value of this book.

I wish to thank all of my colleagues—industry and research partners—who provided insight into the phenomenon of portfolio, programme, and project management.

I would also like to thank Marje Pollack (DerryField Publishing Services), who performed the roles of language editor, copy editor, and typesetter. The quality of this book can be solely attributed to her.

About the Author

Carl Marnewick, PhD, is a professor at the University of Johannesburg, South Africa. He traded his professional career as an information technology (IT) project manager for that of an academic career. This career change provided him with the opportunity to immerse himself in the question of why IT/IS–related projects are perceived as failures and do not provide the originally intended benefits. This is currently an international problem because valuable resources are wasted on projects and programmes that do not add value to the strategic objectives of the organisation.

There is an international problem when there is a gap between theory and practice in organisations that have a presence throughout the world, and Carl is in the ideal position to address this problem because he does practical research within these companies to address this gap. The focus of his research is the overarching concept of the strategic alignment of projects to the vision of the organisations. This alignment forms at the initiation of a project and continues up to the realisation of benefits.

Carl developed a framework—Vision-to-Project (V2P)—which ensures that projects within an organisation are linked to the organisation's vision. Within this framework, a natural outflow of research signifies the realisation of benefits to the organisation through the implementation of IT/IS systems. Benefits realisation results as part of a complex system, and, to date, Carl's research has identified the following impediments to the realisation of benefits: (i) IT project success rates as well as IT project management maturity levels have not improved over the last decade, and these results are in line with similar international research; (ii) IT project managers are not necessarily following best practices and industry standards; (iii) governance and auditing structures are not in place; and (iv) IT project managers' training and required skills are not aligned. If these four aspects are addressed through research and practice, then benefits realisation can occur.

Carl's research has given him a national and international presence. He was actively involved in the development of new international project management standards—that is, ISO 21500 and ISO 21503 (portfolio management)—as well as PMI's *The Standard for Program Management* (4th ed.). Project Management South Africa (PMSA) awarded him the Excellence in Research Award as recognition for his active contribution to the local and global body of knowledge by conducting and publishing scientific research in portfolio, programme, and project management.

Carl is currently heading the Information Technology Project Management Knowledge and Wisdom Research Cluster. This research cluster focuses on research on IT project management and includes, among other topics, governance, auditing and assurance, complexity, IT project success, benefits management, sustainability, and Agile project management.

Chapter 1

Introduction

~ Begin, be bold, and venture to be wise. ~

— Horace*

Executives in any organisation have the duty and obligation to ensure the success of that organisation. Success can be measured in many different ways–for instance, profit or percentage of the market share. The ultimate success is achieved when the vision and organisational strategies are implemented and realised. There are different ways and manners to achieve this, and one arrow in the quiver that will assist is project management (PM).

The question that springs to mind is whether executives should only know how to use the arrow or whether they should know how the arrow is made as well. Executives should not necessarily know the intricacies of project management, but they should know how project management, as a discipline, can benefit the organisation in implementing its strategies and realising its vision. The only way that executives can effectively apply project management to realise these goals is to have sound knowledge of the project management discipline. The purpose of this book is to provide executives with a comprehensive overview of the discipline of project management. The aim is not to explain how to manage projects, as there are a plethora of books that do that. Rather, the book focuses on the benefits of project management to the organisation. The ultimate goal is to provide executives with a view as to how project management can deliver the organisational strategies. The various chapters focus on specific aspects within the project management discipline and how each aspect

* Retrieved from https://www.brainyquote.com/quotes/quotes/h/horace152482.html

should be managed from a business perspective and not necessarily from a project management perspective.

The book covers the entire spectrum of project management from a management and leadership perspective. The focus is not necessarily on what needs to be done from a project management perspective, but on what organisations and senior executives can do to make the facilitation of projects easier.

The first part of the book covers two aspects. The first aspect is what value project management (Chapter 2) has for the organisation at large. The focus is on value-driven project management and the rationale for project management. The second aspect focuses on project management as a strategic enabler (Chapter 3).

The second part focuses on project management *per se* and provides a view of project management from a management perspective. The first concept is portfolio management (Chapter 4) and what it all entails. It then evolves into programme (Chapter 5) and project management (Chapter 6) and how these two disciplines contribute to the success of the organisation at large. These three concepts are explained from the viewpoint of the executive leadership. The fourth concept focuses on the role of the project management office (Chapter 7) as an organisational institution that can facilitate the successful delivery of projects, programmes, and portfolios.

The third part focuses on benefits management (Chapter 8) and sustainability (Chapter 9). The focus is on the benefits that project deliverables bring to the organisation as well as the ultimate sustainability of the organisation. Sustainability within the project is also addressed.

The fourth part focuses on governance within the project management domain (Chapter 10). The focus is on the adherence of projects, programmes, and portfolios to inherent governance principles as well as formal governance aspects, such as legislation. The role of the sponsor is also addressed as a topic for discussion within the realm of governance (Chapter 11).

The layout of the book follows a logical trail, starting with a look at the value of project management. The trail finally ends with the role that executives or organisational leaders can play as executive project sponsors. The book can be read either from start to finish, or each chapter can be read individually. It all depends on the needs of the reader.

Chapter 2 focuses on the value of project management and answers the question of why organisation leaders should invest in project management. Three components have been identified that can be used to determine the value of project management—that is, project management, organisational, and value context. The success rates of projects are briefly discussed, arguing the fact that value cannot be realised given the high number of project failures. The chapter concludes with a section on how executives can contribute to realise the value of project management.

Chapter 3 introduces the notion of the strategic alignment of programmes and projects. The argument is made that all programmes and projects should be aligned to the organisational vision and strategies. A framework is provided that organisational leaders can use to derive programmes and projects from the vision and strategies. The chapter concludes that there are various actions that executives can perform to ensure the strategic alignment of programmes and projects.

Chapter 4 introduces the concept of portfolio management. Current standards on portfolio management are discussed, with an explanation of the portfolio management process. Portfolio management success is explained in the next section, followed by the role of the portfolio manager. As in the previous chapters, this chapter concludes with the executive's role in portfolio management.

Chapter 5 focuses on programme management. The notion of programme management is debated as well as the rationale for a programme. A distinction is made between a programme and a large project, as different skills are needed for each. The same layout is followed as in the portfolio management chapter, where a framework is provided followed by a discussion on the current programme management standards. The role of the programme manager is followed by a brief discussion on what constitutes programme management success. The chapter concludes with the executive's role in programme management.

Chapter 6 focuses on project management and introduces the project management life cycle as well as the process groups. Three standards and one methodology are discussed. The competencies that project managers should master are discussed in the fourth section of this chapter. The focus is on the various competency models and how a competent project manager can have a positive impact on the success of a project. PM maturity is introduced, and the benefits to the organisation when PM as a discipline is mature are explained. The notion of PM success is discussed based on a continuum.

Chapter 7 introduces the concept of the project management office (PMO). The role of the PMO is explained based on five activities that the PMO should perform. The role of the PMO is based on the maturity of the PMO itself, and five maturity stages are explained to position the role of the PMO. The fourth section of this chapter elaborates on metrics and how the PMO should go about determining and measuring metrics that can be used to ensure programme and project success. The various challenges that a PMO faces are discussed, focusing on the role of the executive in the success of the PMO.

Chapter 8 introduces the concept of benefits realisation management. Projects are implemented to generate benefits, and the process of realising benefits is discussed and explained in this chapter. Executives as well as project managers have a specific role to play in ensuring the optimum realisation of the benefits. These roles are unpacked to indicate to executives their role in the benefits realisation process. This chapter continues by discussing a benefit profile and what constitutes a benefit. This information is used to create a benefits dependency network. This network is used to manage benefits throughout the life cycle. The chapter concludes with a discussion on benefits realisation management maturity and the executive's role in benefits realisation.

Chapter 9 addresses the important notion of sustainability. Sustainability is the new buzzword, and executives need to understand how project management and sustainability should be combined. Sustainability is discussed based on the three sustainability dimensions—that is, people, plant, and profit. The focus is on how these dimensions can be introduced into the daily management of a project and contribute to the ultimate sustainability of the organisation itself. A project management sustainability model is provided that executives can use to incorporate sustainability into project management.

Chapter 10 talks to the important notion of governance. Three concepts related to governance are discussed. These concepts are governance of projects, project governance, and governmentality. The positive relationship between project governance and project success is further elaborated on, with a focus on the various roles and responsibilities. A project governance framework provides insights into the different domains and functions. This chapter also focuses on the auditing of projects and the important role of continuous auditing throughout the project's life cycle.

Chapter 11 shifts the focus towards the executives and the role that they play as sponsors of programmes and projects. Attention is given to the characteristics of an executive sponsor and the role that the sponsor should fulfil. Challenges facing an executive sponsor are mentioned as well as how an executive sponsor can be empowered to play a meaningful role.

The ***final chapter*** (Chapter 12) concludes with a comprehensive portfolio, programme, and project management framework. This framework ties all of the concepts together and illustrates the relationships between the various concepts, as discussed in this book. This framework underlines the notion that each concept influences the other, and to achieve value from project management, a comprehensive or holistic view is needed.

Chapter 2

Value of Project Management

~ Price is what you pay. Value is what you get. ~

— Warren Buffet, American Investment Entrepreneur*

Organisational leaders will only invest in something if there is a perception that they will get some value from this specific investment. This is regardless of the type of organisation or the type of investment. For organisations to invest in the project management (PM) discipline, project management should be able to provide benefits back to the organisation. The following question is thus raised: Does project management provide value to organisations? Or if we state it differently: Can organisational leaders function and perform without project management?

Before any argument or conversation can start on the concept of value, it is best to first understand what value means and, more importantly, implies. The *Oxford Dictionary of English* (*ODE*) (Stevenson, 2010) stipulates that value can be either a noun or a verb. When the noun "value" is applied to project management, then the following can be derived from the *ODE*'s definition:

1. Project management must be worth something to the organisation or must be desirable to the organisation.
2. Organisations should be able to exchange project management for either money or goods in the open market.
3. Project management should also be able to substitute or replace something else within the organisation that holds similar value.

---

* Retrieved from https://www.brainyquote.com/quotes/quotes/w/warrenbuff149692.html

4. Project management should be perceived as something well worth the money spent on it.
5. Project management should be able to serve a purpose or cause an effect.

Given the explanation in points one to five, it seems as if pressure is placed on project management to fulfil the expectations associated with the noun "value." At the end of the day, organisations spend some money and other resources on project management, which determines project management's financial value. Second, organisations then, in return, expect that project management will be a solid investment that will bring about either tangible (money, revenue) or intangible benefits. Third, organisations also expect project management to have a positive effect within the organisation—for instance, the improvement of processes such as the strategic alignment of objectives or risk management viewing risks as both threats and opportunities.

2.1 Value Components

Senior executives have queried the value of project management since its inception. Kwak and Ibbs (2000) mention in their study that managers had trouble

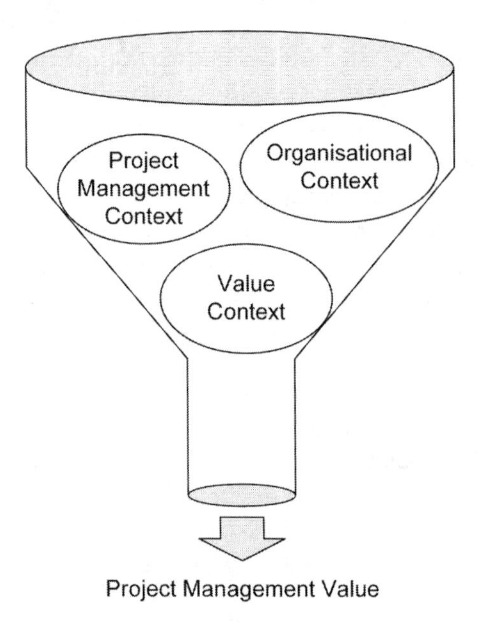

Figure 2-1 Conceptual Model to Determine the Value of Project Management (Adapted from Thomas and Mullaly [2008])

convincing senior executives that their investments in project management actually result in financial and organisational benefits. This viewpoint, wherein the value of project management is queried on a continuous basis, sparked some in-depth study in this regard by the Project Management Institute (PMI). This research was headed by Janice Thomas and Mark Mullaly.

Thomas and Mullaly (2008) proposed a conceptual model to determine the value of project management. The conceptual model consisted of various constructs and highlighted the complexity that is associated with the value of project management. The conceptual model is replicated in Figure 2-1.

The three components can be described as follows:

1. ***Organisational context:*** This is the environment that an organisation is operating in, which has an influence on the value that the organisation derives from project management. The aspects that need to be considered are the culture of the organisation, the economic environment, and the strategic focus, as well as the demographics of the organisation.
2. ***Project management context:*** The aspects that play a role in this component are the organisation's history of implementing project management, the support and structures in place to support project management, and the skills and competence of the project managers themselves.
3. ***Value context:*** This aspect needs to be identified, whereby the value of satisfaction, alignment of business processes, project success and return on investment (ROI) are quantitatively measured.

It is evident that the value of project management is more complex than one might have thought, and the following sections discuss these three components in more detail.

2.1.1 Organisational Context

The value that organisations derive from their investment in project management is directly dependent upon the fit between project management implementation and the context of project management within the organisation (Crawford & Helm, 2009). Project management can and will only provide value to an organisation if the concept of project management is accepted and embraced. It is of no use to implement project management when the organisation itself is of the view that it is a waste of time and energy. This said, it implies that senior executives must create a culture within the organisation wherein project management as a discipline is cherished and promoted. In the same vain, Mengel, Cowan-Sahadath, and Follert (2009) are of the opinion that when project management is supported by the organisation, it then, in return, contributes to the organisational culture and to the larger culture of the overall organisation. It is thus a two-way street, where project management benefits from organisational support, but then in return, an organisational culture is created where the principles of project management are nurtured and embraced.

One aspect of the organisational context is whether the organisation is a public or private organisation. In their study, Crawford and Helm (2009) determined that even for public organisations or entities, value can be determined from the proper use of project management. The results from their study highlighted that project management is seen as assisting governmental entities with the processes and documentation necessary to satisfy auditors. This relates to senior executives playing the role of project sponsors for the various project initiatives. Within the governmental context, the value that these entities derive from embracing project management consists of the improvement of service delivery as well as client and user satisfaction (Crawford & Helm, 2009).

Organisations and senior executives should provide an environment that is conducive to project management and facilitate an environment conducive to project manager performance. The organisational context enables project managers to perform, but it is dependent on the context of project management within the larger organisational context.

2.1.2 Project Management Context

Although the organisational context plays a vital supporting role, the context within which project management functions is important. Aspects that play a role within this context are the organisational structures such as the project management office (PMO); the standards and best practices that the organisation adheres to; and the competence and certification of the portfolio, programme, or project (P3) managers.

One of the benefits of adhering to or following an industry standard or best practices is that it allows the organisation to follow standardised processes. These standardised processes allow the organisation to then unleash the inherent value associated with this standardisation. Therefore, the ability to consistently and predictably deliver results is improved (Crawford & Helm, 2009). This is only possible when the P3 managers are adhering to these standards and do not deviate from or ignore some of the processes. It is important that these standards are adhered to as closely as possible. One of the challenges that P3 managers face within the organisation, as per Kwak and Ibbs (2000), is to implement project management best practices and processes within their respective organisations. They have to prove to the organisation that by following these standards and best practices, the organisation will receive value from the investment in project management. Research by Patah and de Carvalho (2007) determined that organisational leaders who implemented project management saw an improvement of 9.08% in operating margins, which in turn generate an average improvement of US$4.5 million and a total improvement of US$22.3 million. Mengel et al. (2009) support this view, as their research found that the value of project management is directly related to the satisfaction of stakeholders with the management of projects and with project management implementation.

Standards and best practices can only contribute so much to determine the value of project management. These standards and best practices need to be implemented by P3 managers who are competent within the disciplines of project, programme, and portfolio management. As many as possible of an organisation's P3 managers must be certified to improve their competence levels. Competence levels will also be improved through the practical implementation of projects. This implies that senior executives must create an organisation that is conducive for project management. In the PMI's 2015 *Pulse of the Profession®* study, they determine that when a P3 management mindset is embedded into an organisation's DNA, performance improves and competitive advantage accelerates (Project Management Institute, 2015).

One of the vehicles to embed P3 management into the organisation is through the establishment of a PMO. The functions of the PMO include, among others, the following: (i) attention to ensuring consistency and uniformity in projects; (ii) an organisational desire to excel, (iii) an enterprise focus on the improvement in project management competency; (iv) an enterprise focus on knowledge management; (v) a reduction in project overruns; (vi) an increase in the delivery speed of projects; and (vii) an increase in customer satisfaction. Aubry, Hobbs, and Thuillier (2009) proposed that a PMO should not be considered as an isolated island in the organisation but rather as one part of an archipelago that is the organisation at large. They also suggest that project management *per se* and specifically the PMO should evolve continuously. The PMO should adapt to changes in its external or internal environment or as an answer to internal tensions. The PMO is discussed in detail in Chapter 7.

2.1.3 Value Context

The value context focuses on the soft issues such as the satisfaction with the project management process, alignment of business processes and project management processes, as well as project success. All of these need to be quantitatively valued in order for senior executives to see the value of project management. Unfortunately, organisations are managed and run on a monetary basis, and therefore everything needs to be measured against profit and loss. This makes it extremely difficult because how do you determine the monetary value of process alignment? At the end of the day, the ultimate driver is the delivery of benefits from project management and not the associated costs (Crawford & Helm, 2009).

One of the concepts within the value context that Crawford and Helm (2009) determined is that project management has been beneficial for staff morale and satisfaction. The reason for this can be attributed to various factors, including (i) improved customer relations; (ii) shorter development times; (iii) higher quality and increased reliability; and (iv) improved productivity. This increase in staff morale and satisfaction is achieved even against a background wherein project management standards and methodologies are too time consuming and bureaucratic (Crawford & Helm, 2009).

Project management also contributes value to the overall accountability and transparency associated with governance. Value is derived from the provision of evidence to the various stakeholders. Evidence includes, among others, levels of compliance, risk management, change management, improvement in the organisational culture, and the overall improvement of performance (Crawford & Helm, 2009; Mengel et al., 2009). As indicated within the project management context, adherence to standards is an important aspect of value delivery. Compliance with the standards provides additional value to the organisation in terms of audit and review. Project managers and the project management process can be audited against the standards and deviations and can be reviewed and rectified.

The argument is that organisations should and do get value from their investments in project management based on the organisational, project management, and value contexts. It might be that some organisations derive more value from one context than from another context. The value of project management cannot be divided into three equal parts. The value is really dependent on the organisation itself and where the current focus is with regard to value generation from the project management investment.

The next section briefly discusses the current state of project success. The aim is not to go into a philosophical debate about what constitutes project success. The aim is rather to provide current statistics with regard to project success and to determine whether organisations can still claim value when projects are delivered late and at an increased cost.

2.2 Project Success Rates

Project success has been researched by various researchers and institutions across the world. The research focuses on what constitutes project success and what is meant by this term. The research also focuses on what the perception is of organisations and individuals with regard to project success. In other words, how successful their projects are at the end of the day.

2.2.1 Definition of Project Success

It is important that the concept of project success be defined in general as well as specifically for an organisation within a specific industry. Unfortunately, current literature remains vague regarding project success. Project success is defined by the Project Management Institute (2013) as the quality of the product and project, timeliness, budget compliance, and the degree of customer satisfaction. According to the P2M, "a project is successfully completed [when] it delivers novelty, differentiation and innovation on its product, either in a physical or service form" (Ohara, 2005, p. 16). The *APM Body of Knowledge* defines project success in a similar fashion, stating that it is the project stakeholders' needs that must be satisfied, and this is measured by the success criteria as identified and agreed upon at the start of the project (Association for Project Management, 2006). PRINCE2®, on the other hand, does not explicitly define project success but states that the objectives of the project need to be achieved (Office of Government Commerce [OGC], 2009). Table 2-1 provides the generic criteria of project success as per the various standards and methodologies.

In order to understand project success in its totality, Hyväri (2006) suggests that the critical success factors (CSFs) must also be determined. If the CSFs are in place, then project success should follow as a natural outflow. Hyväri (2006) suggests five CSFs, and each of these has sub-CSFs. These CSFs range from the project itself (size and end-user commitment) to the environment (competitors, nature, and social environment). If companies and project managers focus on these CSFs, then project success should be assured.

Table 2-1 Project Success Criteria

Project Success Criteria	Standard or Methodology
Innovation	Project Management Association of Japan: P2M
Quality	PMI: *PMBOK® Guide*
Timeliness	PMI: *PMBOK® Guide*
Budget compliance	PMI: *PMBOK® Guide*
Customer satisfaction	PMI: *PMBOK® Guide* APM: *APM Body of Knowledge*
Objectives	OGC: PRINCE2®
Novelty	Project Management Association of Japan: P2M
Differentiation	Project Management Association of Japan: P2M

Projects and their subsequent products and/or services cannot be seen in isolation. According to Bannerman (2008), the success of the project can be measured on five levels: (i) process; (ii) project management; (iii) product; (iv) business; and (v) strategy. A project might deliver a product or service late and over budget, but it still delivers on the company's strategy. Is the project then a failure or a success? This multilevel view is supported by Thomas and Fernández (2008), who focus not on five, but three levels—that is, project management, technical, and business.

Ika (2009) investigated the concept of project success and concluded that there was a shift from measuring projects only according to time, cost, and quality to a more holistic approach, which includes the original iron triangle but also focuses on benefits to stakeholders and the organisation at large. The literature review indicates that there are various means that can be used to measure project success.

2.2.2 Project Success Rates

Various longitudinal studies, such as the Chaos Chronicles and Prosperus Reports as well as once-off studies, indicate that everything is not well when it comes to the successful delivery of projects.

One of the most important studies on project success is that of The Standish Group. They have been determining the success of projects for almost two decades and publish the results as the Chaos Chronicles (Eveleens & Verhoef, 2010; The Standish Group, 2013, 2014). The focus of these reports is within the Information Technology (IT) discipline and covers IT projects within the United States and Europe. The results from these studies since 2011 are depicted in Figure 2-2.

The results indicate two important aspects. The first is that many IT projects are not successful. On average, only 30% of the projects are successful, with close to 20% of IT projects perceived as failures. The second aspect is that the success rates are stagnant and there is no improvement with regard to project success.

IT project success from a South African perspective also does not look much better, as depicted in Figure 2-3. A longitudinal study that is conducted regularly by the University of Johannesburg in South Africa indicates that IT project success is comparable with the results of the Chaos Chronicles (Labuschagne & Marnewick, 2009; Marnewick, 2013; Sonnekus & Labuschagne, 2003).

Even in South Africa, it is evident that IT project success is stagnant and that there is no improvement. IT projects keep on failing, which ultimately raises the question of whether executives do receive value when it comes to investing in IT projects.

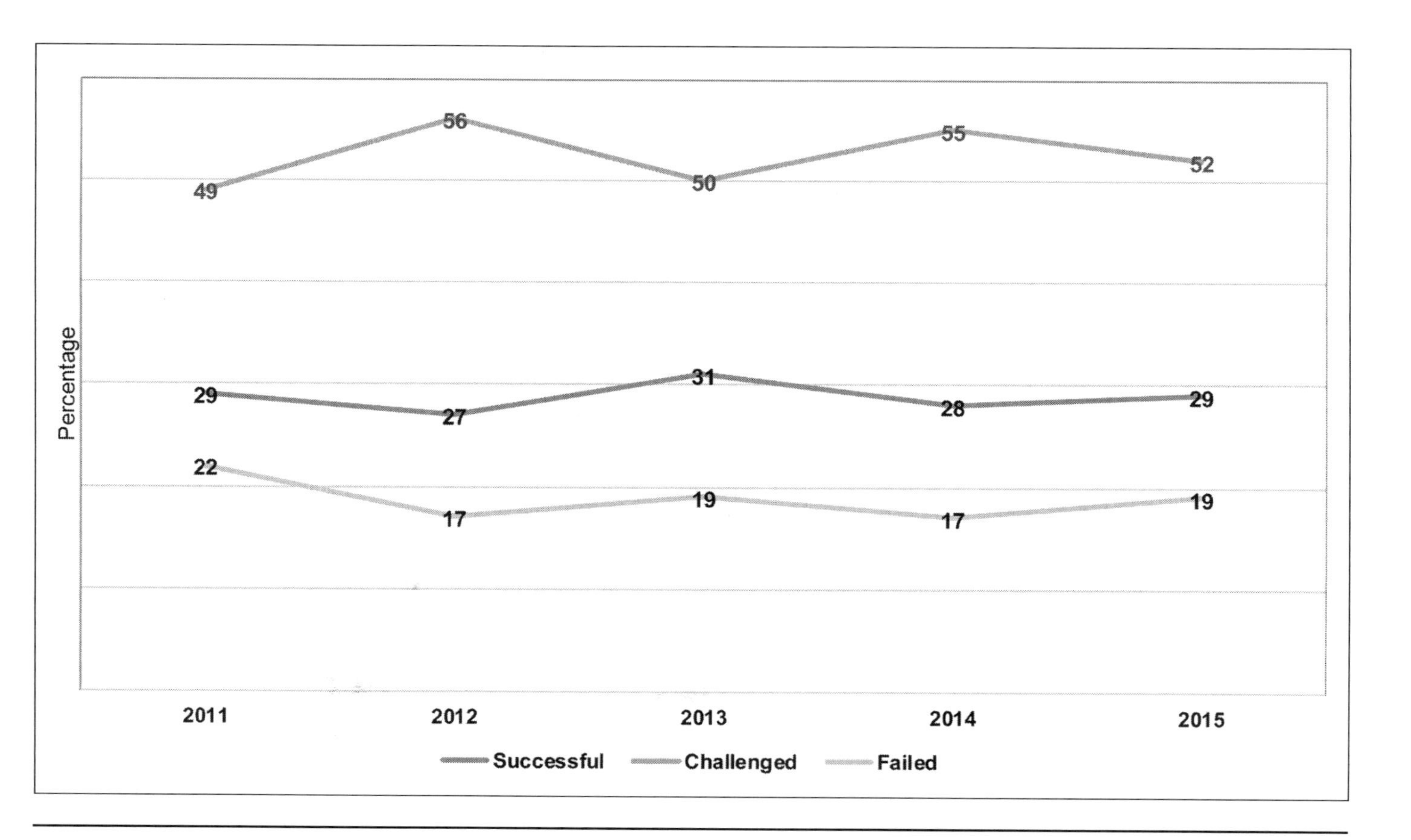

Figure 2-2 Project Success (2011–2015)

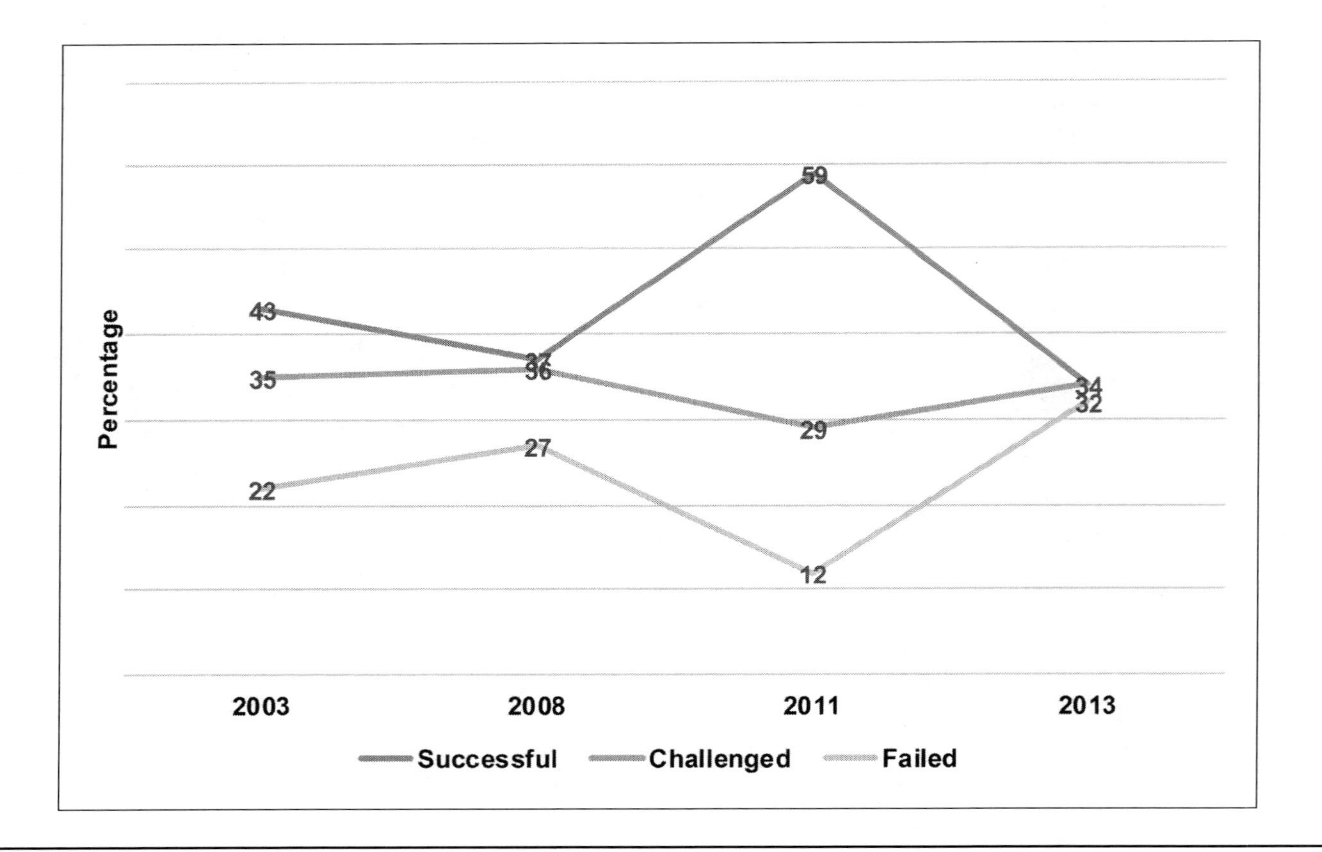

Figure 2-3 South African IT Project Performance Rates

The argument made is that IT projects are notorious for being late, over budget, and not delivering value to the organisation. A similar study on construction and engineering projects within South Africa yields a better picture, but even in this industry the question can be asked whether projects deliver value. The results in Figure 2-4 indicate that close to 50% of construction and engineering projects are successful, but it still leaves half of all the projects as either failed or challenged projects.

The statistics tell a concerning story, and senior executives do have the right to ask whether project management *per se* is the correct vehicle to implement initiatives. The counter argument regards whether another process or management tool is available to replace project management.

The answer lies in what Zhai, Xin, and Cheng (2009) perceive as value. They divide value into two concepts: the value of a project on the one side and the value of project management on the other side. First of all, the value of a project has two main characteristics that focus on the multidimensionality of a project; and second, it focuses on the dynamics within a project. The purpose of project management *per se* is to realise the value of a project. The value of project management includes two dimensions (Zhai et al., 2009). First of all, project

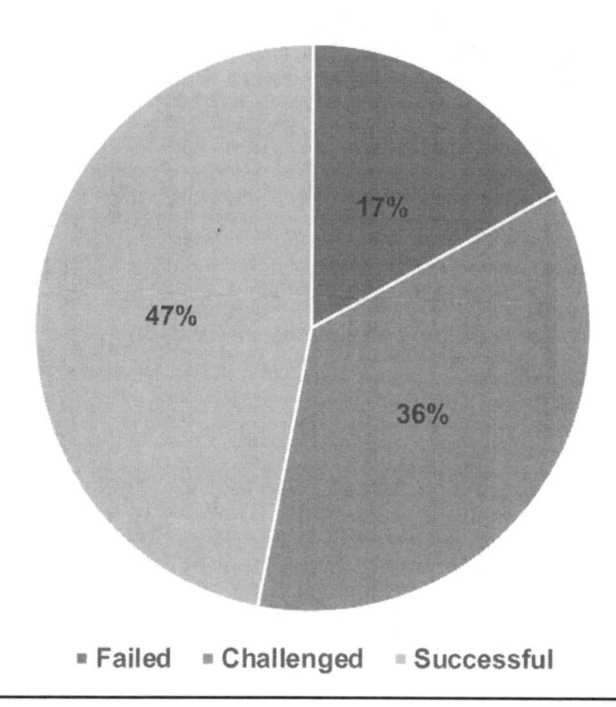

Figure 2-4 South African Construction and Engineering Project Performance Rates (2011)

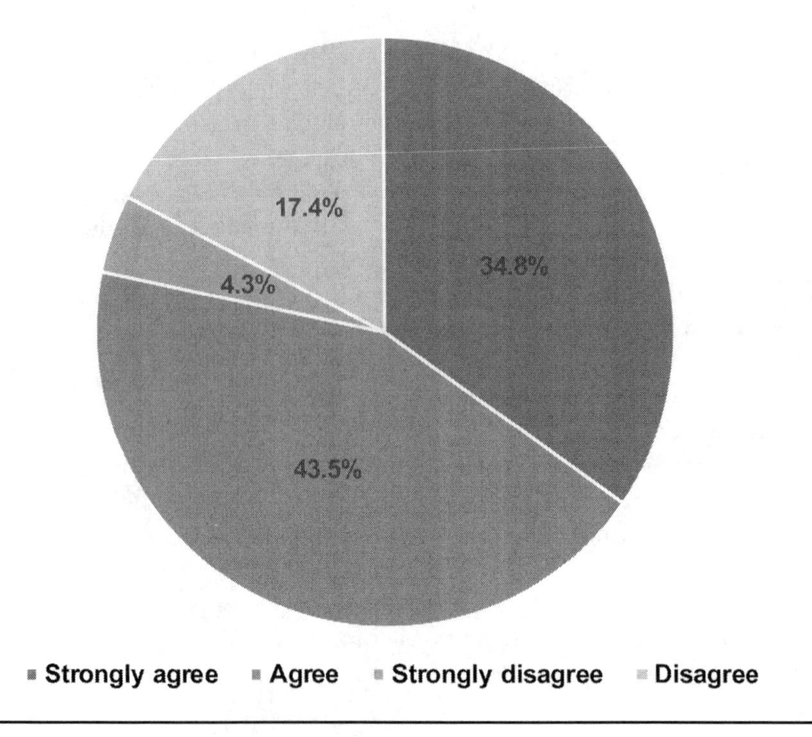

Figure 2-5 Level of Business Value Achieved

management realises the value of a project through the conversion of resources into the project outputs. Second, the value of project management comprises the sum of all the values that stakeholders perceive, including aspects such as cost saving, performance improvement, and other interests through the process of value realisation of the project. In a separate study done by Chalale (2016), he found that business projects are on average 131% over budget, and these projects take, on average, 71% longer than estimated. Of these projects that are late and over budget, only 74% deliver the full scope or requirements. What is remarkable, however, is that regardless of these statistics, the stakeholders do believe that they receive value, as per Figure 2-5.

The results and discussion clearly indicate that the value of project management is determined based on multiple levels, and it is not a single construct.

2.3 The Executive's Role in Realising Value from Project Management

Executives play an important role in realising value from project management. Realising value from project management is a double-edged sword for executives: To realise value from project management, executives must provide an environment conducive to project management. How can executives create a conducive environment? The first thing that executives can do is to determine and establish the enablers for project management.

2.3.1 Project Management Enablers

The first thing that executives must understand is that there is no "one shoe fits all" kind of approach. Mir and Pinnington (2014) indicate that the value of project management is dependent, among others, on the culture of the organisation and the type of implementation "fit" that is in line with the organisation's needs. Some enablers include but are not limited to:

- ***Full project management:*** Executives must support the full implementation of the project management philosophy or discipline within the organisation. Project management cannot be implemented in stages. When a project management standard or methodology is chosen, then it must be implemented and adhered to, to the letter.
- ***Project management office:*** It is advisable that a structure such as a PMO be created. The role of the PMO varies, but ultimately it supports the project manager as well as the organisation at large. Organisations, especially larger ones, cannot really function without a PMO. The role and functions of a PMO are discussed in Chapter 7.
- ***Long-term training:*** Training is essential for the development of any employee. This is especially the case with project management. There are various developments with regard to the tools and techniques that project managers use, as well the continuous updating of standards and methodologies. To get the best performance from project managers, organisations need to continuously invest in the training of these individuals.
- ***Project managers' authority:*** Project managers should be given full authority to manage their respective projects. This is especially difficult within organisations that have a hierarchical or weak-matrix structure. Project managers need to be efficient and able to make quick, informed decisions. This is not possible when the project manager has little or no authority and needs the approval of some executive higher up in the command chain.
- ***Clear roles:*** Parallel to the project manager's authority is the definition of roles and responsibilities. Employees perform best when they know what is expected of them. This is only possible when the roles are clearly defined and when responsibilities, and subsequent accountability, are assigned to these roles. It makes the life of a project manager easier if he or she knows

what the role of each specific individual is within the project team. It eradicates misunderstandings, and the duplication of work is minimised. Project managers should then also have the authority to discipline individuals who are not performing.

- ***Strategic alignment:*** Projects are perceived as the vehicles to implement the organisational vision and strategies. One of the most important enablers that executives can put in place is to ensure that all projects within an organisation are aligned with the organisational vision and strategies. This alignment ultimately creates value, as the vision and strategies are realised once the projects are completed.

The enablers provided here are but a short list, and each organisation must determine what enablers they must put in place to create value from the project deliverable itself and from the project management discipline.

2.3.2 Metrics to Measure Value

In order for organisations to determine whether they are achieving value, certain metrics must be in place. No executive will run and manage a business without daily measures and results to gauge whether the business is on the right track or not. The same applies to the value of project management. Executives must find metrics to determine the value of the project deliverable itself and the value of project management. Bannerman (2008) created a multilevel framework that can be used to determine the value created. This framework is displayed in Figure 2-6.

Executives must ensure that each and every project within the organisation is measured to determine the value of the project itself as well as the value of the deliverable, which in turn will determine the value of project management. Three components are measured to determine the value—that is, the project itself, the product that is delivered by the project, and the impact of this product on the organisation itself.

1. When one looks at the project component, the first aspect to measure is the value that the process itself contributes to the organisation. The aim is to measure whether discipline-specific technical and managerial processes, methods, tools, and techniques are employed to achieve the project objectives. The second aspect that needs to be measured is that of the constraints within which the project is implemented. The traditional constraints are time, cost, scope, and/or quality. The purpose here is to determine whether the project was delivered within the constraints of the project.

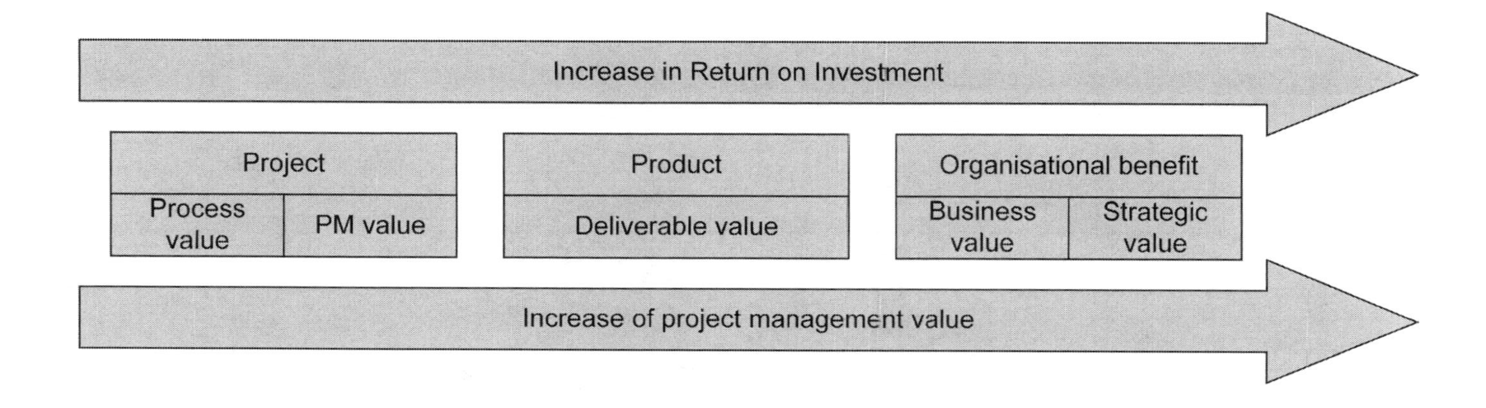

Figure 2-6 Framework to Measure Value Creation (Adapted from Bannerman [2008])

2. The second component is the main deliverable(s) from the project, and the nature of the deliverable(s) will be discipline specific. Measurements that spring to mind are whether (i) the specifications are met; (ii) the requirements are met; (iii) the client/user expectations are met; (iv) the client/user accepted the deliverable(s); (v) the product/system is being used; (vi) the client/user is satisfied; and (vii) the clients' benefits are realised.
3. The third component focuses on the organisational benefits that are associated with the specific project. The first aspect here emphasises the business value itself and the organisational objectives that motivated the investment—that is, what the organisation wanted to achieve from the investment. The second aspect focuses on the strategic value and the strategic advantage gained from the project investment, either sought or emergent. Executives must put metrics in place to measure the organisational benefits such as whether (i) the business objectives were met; (ii) the organisational benefits were realised; (iii) organisational development was enabled; and (iv) positive stakeholder response was generated.

Regardless of which model or framework is used, executives must ensure that each and every project is scrutinised to determine whether the project is delivering value to the organisation. This scrutiny must happen right at the start of the project, when the business case is evaluated; throughout the project life cycle; and after the completion of the project, when benefits are realised. It must be noted that in the case of organisational improvement projects, benefits can be realised throughout the project life cycle. Benefits realisation is discussed is more detail in Chapter 8. It is not an easy task, but once everyone realises that each and every project is scrutinised, value will start to emerge from investments in the various projects.

Value creation from projects and project management is unfortunately not a once-off exercise. Executives must ensure that the value of project management is sustained.

2.3.3 Sustainability of Value

To sustain the value of project management, executives should continuously reinvest in the enablers, as discussed in Section 2.3.1. This reinvestment in the enablers then provides, in return, the various contexts, which ultimately determine the value of project management. These relationships are illustrated in Figure 2-7.

Ultimately, the value of project management can be measured at the following levels (Zhai et al., 2009):

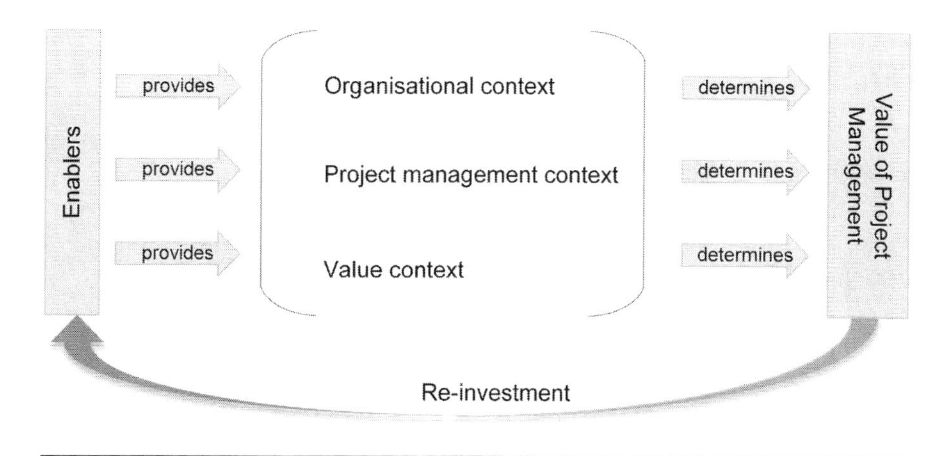

Figure 2-7 Sustainment of Value of Project Management

1. ***Stakeholder satisfaction:*** Stakeholders need to be satisfied with the project deliverables, and this can be measured according to the framework depicted in Figure 2-6.
2. ***Aligned use of practices:*** Project management falls within the larger organisation, and, therefore, the best practices employed by project management must align with all of the organisation's other business processes.
3. ***Business outcome:*** Value is created when the project deliverables complement the realisation of the organisational vision and strategies.
4. ***Return on Investment (ROI):*** ROI ultimately links the stakeholders to the value of project management—where value covers tangible financial returns as well as intangible organisational benefits.

2.4 Conclusion

Can organisations perform and deliver on the vision and strategies without project management? It is most likely possible, but organisations stand a better chance to be successful when they embrace project management as a vehicle to deliver value.

This value creation is not going to happen without any investment from the organisation itself. As we have seen, various enablers need to be in place, and the only ones that can put these enablers in place are the executives of the organisation. Yes, project managers *per se* must also play their part, but without the enablers, their hands are tied. These enablers then provide the context for project management to provide value to the organisation.

Executives must not be fooled and believe that it is an easy process to create value from project management. It is a continuous process, and through this process, the project management maturity of the organisation improves and so will the ability to deliver projects more successfully. This, in turn, ensures that more value is created from the investment in project management.

The next chapter focuses more specifically on how organisations can derive value from project management through the strategic alignment of project and programme management with the organisational vision and strategies.

2.5 References

Association for Project Management. (2006). *APM Body of Knowledge* (5 ed.). Buckinghamshire, UK: Association for Project Management.

Aubry, M., Hobbs, B., & Thuillier, D. (2009). The Contribution of the Project Management Office to Organisational Performance. *International Journal of Managing Projects in Business, 2*(1), 141–148.

Bannerman, P. L. (2008). *Defining Project Success: A Multilevel Framework*. Paper presented at the PMI Research Conference: Defining the Future of Project Management, Warsaw, Poland.

Chalale, M. L. (2016). *Investigating the Relationship between IT Project Expenditure to IT Project Success and IT Business Value*. (M.Com. Business Management), University of Johannesburg, Johannesburg, South Africa.

Crawford, L. H., & Helm, J. (2009). Government and Governance: The Value of Project Management in the Public Sector. *Project Management Journal, 40*(1), 73:87.

Eveleens, J. L., & Verhoef, C. (2010). The Rise and Fall of the Chaos Report Figures. *IEEE Software, 27*(1), 30–36.

Hyväri, I. (2006). Success of Projects in Different Organizational Conditions. *Project Management Journal, 37*(4), 31–41.

Ika, L. A. (2009). Project Success as a Topic in Project Management Journals. *Project Management Journal, 40*(4), 6–19.

Kwak, Y. H., & Ibbs, W. C. (2000, 15 August). *The Berkeley Project Management Process Maturity Model: Measuring the Value of Project Management*. Paper presented at the Proceedings of the 2000 IEEE Engineering Management Society, Albuquerque, NM, USA.

Labuschagne, L., & Marnewick, C. (2009). *The Prosperus Report 2008. IT Project Management Maturity vs. Project Success in South Africa*. Johannesburg, ZA: Project Management South Africa.

Marnewick, C. (Ed.) (2013). *Prosperus Report—The African Edition*. Johannesburg, ZA: Project Management South Africa.

Mengel, T., Cowan-Sahadath, K., & Follert, F. (2009). The Value of Project Management to Organizations in Canada and Germany, or Do Values Add Value? Five Case Studies. *Project Management Journal, 40*(1), 28:41.

Mir, F. A., & Pinnington, A. H. (2014). Exploring the Value of Project Management: Linking Project Management Performance and Project Success. *International Journal of Project Management, 32*(2), 202–217.

Office of Government Commerce. (2009). *Managing Successful Projects with PRINCE2* (5 ed.). United Kingdom: The Stationary Office.

Ohara, S. (2005). *P2M: A Guidebook of Project & Program Management for Enterprise Innovation* (3 ed.). Tokyo, JP: Project Management Association of Japan (PMAJ).

Patah, L. A., & de Carvalho, M. M. (2007, 5–9 August). *Measuring the Value of Project Management.* Paper presented at the PICMET '07—2007 Portland International Conference on Management of Engineering & Technology, Portland, OR, USA.

Project Management Institute. (2013). *A Guide to the Project Management Body of Knowledge (PMBOK® Guide)* (5 ed.). Newtown Square, PA, USA: Project Management Institute.

Project Management Institute. (2015). *Capturing the Value of Project Management.* Newtown, PA, USA: Project Management Institute.

Sonnekus, R., & Labuschagne, L. (2003). *The Prosperus Report 2003.* Johannesburg, ZA: RAU Standard Bank Academy for Information Technology.

Stevenson, A. (2010). *Oxford Dictionary of English.* Oxford, UK: Oxford University Press.

The Standish Group. (2013). *Chaos Manifesto 2013.* Retrieved from http://www.immagic.com/eLibrary/ARCHIVES/GENERAL/GENREF/S130301C.pdf

The Standish Group. (2014). *Chaos Manifesto 2014*: The Standish Group International, Inc.

Thomas, G., & Fernández, W. (2008). Success in IT Projects: A Matter of Definition? *International Journal of Project Management, 26*(7), 733–742.

Thomas, J., & Mullaly, M. (2008). *Researching the Value of Project Management.* Newtown Square, PA, USA: Project Management Institute.

Zhai, L., Xin, Y., & Cheng, C. (2009). Understanding the Value of Project Management from a Stakeholder's Perspective: Case Study of Mega-Project Management. *Project Management Journal, 40*(1), 99–109.

Chapter 3

Strategic Alignment of Programmes and Projects

- Strategy without tactics is the slowest route to victory, tactics without strategy is the noise before defeat. -

— Sun Tsu, Ancient Chinese Military Strategist*

Decisions are made on a daily basis by different people within the organisation (Van Der Merwe, 2002). These decisions influence the way the organisation is run in the short as well as the long term (Desai, 2000). The decision-making process within an organisation explains and provides a clear understanding of the relationship between the levels of decision making and the influence of these decisions on projects in general. This addresses one of the components within organisational theory, which focuses on various approaches to analyse organisations. Organisational theory can be defined as the way an organisation functions and how it affects and is affected by the environment (Jones, 2001, p. 8). The other two components address the design and culture of the organisation.

Managing activities internal to an organisation is only part of the executive's responsibilities (Nadler, 2004). The executive must also respond to the challenges posed by the organisation's immediate and remote external environments. To deal effectively with everything that affects the growth and profitability of an organisation, executives employ management processes that position the organisation as optimally as possible in its competitive environment by maximising the anticipation of environmental changes and unexpected internal and competitive demands (Kakabadse et al., 2001).

* Retrieved from http://philosiblog.com/2011/05/04/strategy-without-tactics-is/

This all-encompassing approach is known as strategic management, which can be defined as the set of decisions and actions that result in the formulation and implementation of strategies designed to achieve an organisation's objectives (Dew, Goldfarb, & Sarasvathy, 2006; Richardson, 1994). A strategy is a large-scale, future-oriented plan for interacting with the competitive environment to achieve the organisation's objectives (Manas, 2006). Although the plan does not detail all future deployments, it does provide a framework for managerial decisions (Joshi, Kathuria, & Porth, 2003).

3.1 Strategic Alignment of Programmes and Projects

It is widely accepted that projects are used to implement a specific organisational strategy (Cooke-Davies, Crawford, & Lechler, 2009; Hermano & Martín-Cruz, 2016). Furthermore, the types of projects that are initiated and even managed are dependent upon the main strategic focus of the organisation (Cooke-Davies et al., 2009). A review of the literature reveals that there is enough theoretical and practical information on how to formulate organisational strategies and the causal effect of effective strategy implementation on organisational performance. Regardless of this knowledge, Crawford (2014) is of the opinion that strategy implementation *per se* has received less research attention than the formulation of strategies itself. Additionally, in terms of how effective strategies are formulated, it is of no or little value if these strategies are not balanced by effective implementation (Crawford, 2014). She continues the debate and suggests that programmes and projects are the only vehicles for the implementation of strategy and that they position the management of projects as a key business process. Executives should take caution that not all aspects of a strategy are implemented through programmes and projects. Business-as-Usual (BAU) or operations are also part of strategy implementation, and the focus is on the day-to-day achievement of its purposes (Crawford, 2014). These day-to-day activities might be implemented either through projects or through normal activities.

It is important for an executive to note that successful strategy implementation through programmes and projects is directly related to the amount of resource allocation that a project manager controls (Eweje, Turner, & Müller, 2012). A balance needs to be created wherein an executive as well as a project manager control the resources. The purpose is not to create a power struggle but to ensure that the strategies can be implemented effectively. The assumption is that the scale will be tipped toward the executive, as he or she is the more senior person and the more appropriate person to control the necessary resources. This said, it also implies that when an executive as well as a project manager are in control of organisational resources, then the chances are much higher that the strategies will be successfully implemented. There is sometimes a tendency for executives to allow so-called "pet" projects to flourish. Executives provide resources and motivation to keep these "pet" projects within the portfolio beyond justification (Hermano & Martín-Cruz, 2016). A "pet" project is a project pushed by an executive that adds no value whatsoever to the organisation but may be beneficial to the executive's own personal and secret agenda. "Pet" projects usually drain the organisation's resources (monetary and human resources) for nothing, with the exception of the satisfaction of the executive.

One of the important aspects of the strategic alignment of programmes and projects is the strategic value that is created and released through this alignment. Executives should realise that since projects enable the successful delivery of long-term strategies, they should find ways to unleash strategic value through the alignment of programmes and projects (Martinsuo, Gemünden, & Huemann, 2012). The alignment of programmes and projects goes beyond pure alignment and strategy implementation. Organisations must identify and measure the strategic value that is unleashed through the implementation of strategically aligned programmes and projects (Martinsuo & Killen, 2014). Within the portfolio, strategic value should be used to balance the importance and range of strategies. This then implies that strategic value will also be used to balance the portfolio to determine which programmes and projects should be implemented to achieve the maximum financial as well as strategic value. Executives should be warned that the immediate financial success of a portfolio should be separated from measuring overall business success as business success is measured in terms of both economic value and in preparing the organisation for the future.

Although projects are increasingly being undertaken to implement business strategy, it is not always that easy to determine whether these projects are contributing to the realisation of strategic goals (Young & Grant, 2015). Organisations should have metrics and measurements in place to determine whether, at an operational and project level, there is an improvement (Young & Grant, 2015).

Young, Young, Jordan, and O'Connor (2012) state that although there is significant evidence for the value of aligning programmes and projects to organisational strategies, there is little guidance about how strategy gets translated into projects. The next section provides a framework that can be used to derive programmes and projects from the vision and strategies of the organisation.

3.2 Aligning Projects with the Organisational Vision

Although the *Organizational Project Management Maturity Model* (OPM3®) (Project Management Institute, 2009) recognises the fact that the vision and strategies of an organisation are implemented by means of projects, it does

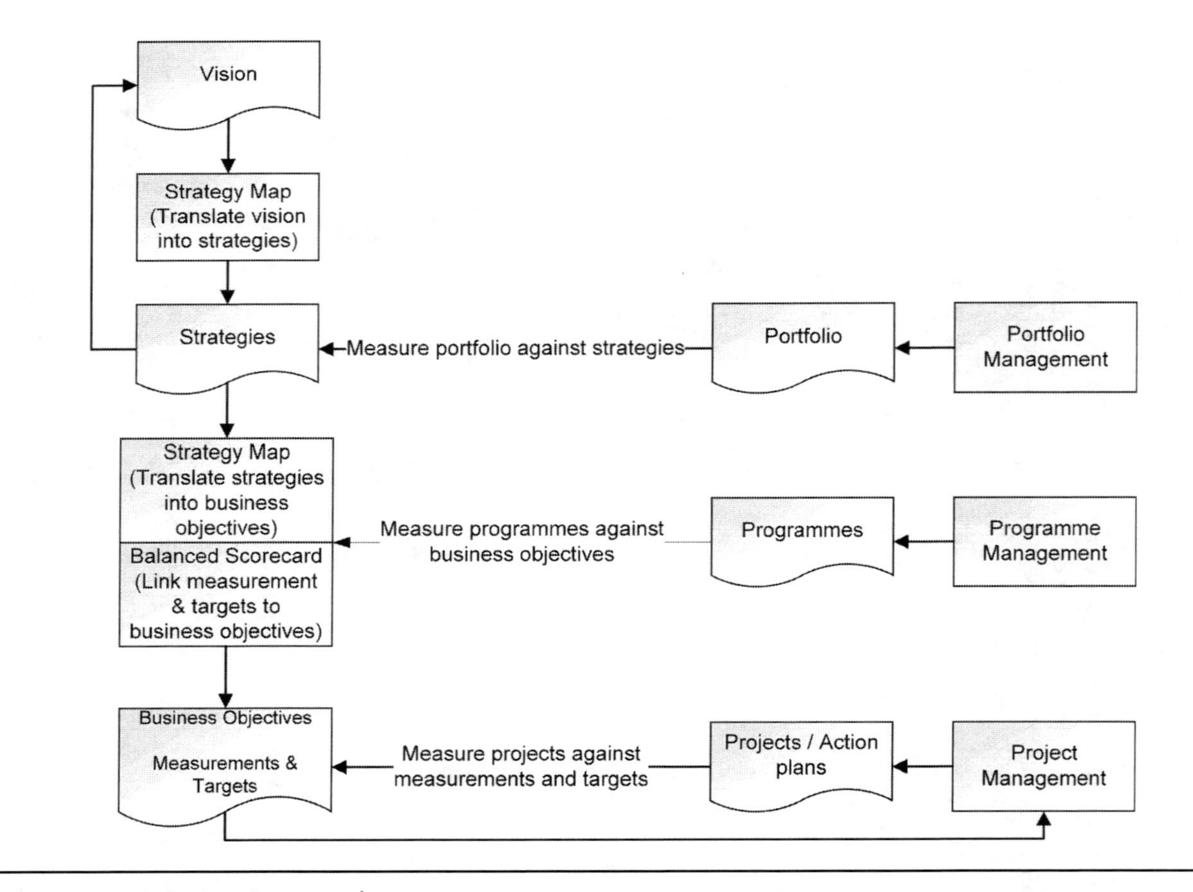

Figure 3-1 Vision-to-Project Framework

not provide a clear approach for proceeding from the vision to the projects. According to PriceWaterhouseCoopers, "any project undertaken by a company should be driven by business objectives" (Peterson, 2002, p. 1). Many organisations lack a structured process through which to derive projects from their business objectives. Longman and Mullins (2004) also acknowledge the fact that an organisation's strategy should provide the boundaries for programmes and projects.

The question is: How should a project portfolio be constructed and managed in order to implement and fulfil the organisational strategies?

Figure 3-1 suggests a structured way and manner that might be useful in deriving projects from the organisational vision and continuously monitoring the contribution that these projects make toward achieving the vision (Marnewick & Labuschagne, 2006).

The following sections discuss each of the components of the Vision-to-Project (V2P) Framework.

3.2.1 Vision

A vision is a concrete idea that describes what needs to be achieved by the organisation as well as how it should be done. Formulating a vision is a complex process, and it is important that the vision be articulated explicitly, and that it generates enthusiasm (Ezingeard, McFadzean, & Birchall, 2007; Testa, 1999). Most important is the necessity that the vision must be realistic and credible, and that it projects a future attractive enough to convince the followers to invest efforts in pursuing it, rather than simply continuing with the status quo. The vision should be based on reality and should not be a far-fetched dream (Bogler & Nir, 2001; Pearce & Robinson, 2000). It is the duty of executives to quantify the vision. Different methods are available, such as McKinsey's 7-S framework (Rasiel & Friga, 2001) or Kenichi Ohmae's 3Cs framework (Ohmae, 2001). McKinsey's 7-S framework is used as an analysis tool to assess and monitor the organisation itself. The focus is on seven elements—structure, strategy, systems, skills, style, staff, and shared values. The idea is that these elements should be aligned in order for organisations to perform. The 3Cs framework focuses on customers, competitors, and the corporation. These three elements should be addressed by organisational leaders in order for the organisation to succeed.

The problem with these methods is that they provide ways and means to determine and describe strategies. Neither of these methods provides a methodology or a process to link projects to the vision and strategies of the organisation. The combination of strategy maps and balanced scorecards provides a process whereby an organisation can link projects to the vision and strategies.

Strategy maps are used to derive business objectives from the vision. The balanced scorecard is then used to derive measurements for each strategy. This allows an organisation to measure the progress of a strategy.

3.2.2 Strategy Maps—Translating a Vision into Strategies

Strategy maps describe the vision and strategies of the organisation by means of processes and intangible assets (Kaplan & Norton, 2004; Marr & Adams, 2004). Strategy maps can be used to align intangible assets with the organisational strategies and, ultimately, the vision of the organisation. A strategy map starts with a vision and follows an initial top-to-bottom and then a bottom-to-top approach. This approach means that the vision dictates all the lower levels, but the bottom-up approach enables the organisation to link everything back to the vision. Although the approach is a top-to-bottom approach, the lower levels must be linked to the upper levels to ensure that there is a consistent link between the upper and lower levels of the strategy map.

The strategy map provides a framework for an organisation, but it is up to the organisation to determine which of the components it wants to use. The four perspectives—financial, customer, internal processes, and learning and growth—provide the different focal points of the strategy map. The financial and customer perspectives can be used to derive the strategies, whereas the internal and learning and growth perspectives provide the business objectives.

The first perspective is the financial perspective, which provides the two main strategies for the organisation—namely, the growth and productivity strategies. This allows an organisation to be focused on specific strategies and not spread its attention to different strategies that do not facilitate the implementation of the vision. The executives of an organisation should find a balance between the growth and productivity strategies. These two strategies are often in direct conflict, and it is crucial to the organisation that a balance is kept. The financial strategies focus on how the organisation can increase profitability and return on investment (ROI) (Rowley, 2002; Thackray, 1995). This is not possible without a strategy that links this to the needs of prospective customers.

The customer perspective focuses on this aspect of the organisation, and the organisation identifies the targeted customer segments in which it wants to compete. The strategy map of an organisation produces financial and customer strategies that are linked to the vision of the organisation. The realisation of the strategies ultimately ensures the success of the organisation as they are linked to the organisational vision.

3.2.3 Strategy Maps—Translating Strategies into Business Objectives

Once the strategies are determined, the third and fourth perspectives of the strategy map can be applied to determine the business objectives. The business objectives focus on the internal processes of the organisation as well as the learning and growth perspective. Managers identify the processes that are most critical for achieving customer and financial strategies. This enables an organisation to focus the internal business processes on those processes that will deliver the determined strategies in the most successful manner. Two processes might address the same business objective, but one process is more optimal than the other and is, therefore, a better solution (Kaplan & Norton, 1996). Once the internal processes of the organisation are determined, they must be linked to the intangible assets of the organisation. This is done through the learning and growth perspective. This perspective, together with the internal perspective, forms the basis of the business objectives. The learning and growth perspective highlights the role for aligning the organisation's intangible assets with its strategy. Intangible assets can be organised into three categories—human, information, and organisational capital. The objectives in these three categories must be aligned with the objectives of the internal processes and integrated with one another. These objectives describe the importance of intangible assets and provide a powerful framework for the management of intangible assets.

3.2.4 Balanced Scorecards—Determining Targets and Measurements

The balanced scorecard provides executives with a comprehensive framework that will translate an organisation's vision and strategies into a coherent set of performance measures (Kaplan & Norton, 2004; Ritter, 2003). The balanced scorecard provides measurement criteria for the business objectives derived from the vision and strategies using a strategy map. Although the different perspectives are tied to a specific level within the strategy map, it must be noted that these perspectives influence the other levels, and there is coherence between the perspectives and the strategies and business objectives of the organisation.

The next step in the process is to take the measurement criteria and targets and translate them into projects. Various projects will be used to implement a product and/or service that will satisfy the measurement criteria.

3.2.5 Project Management

The strategy map in conjunction with the balanced scorecard provides the organisation with business objectives with the relevant measurement criteria and targets to determine the success of an individual business objective.

The principles of project management are applied to each business objective that was defined by the strategy map. This means that a business case will be done to determine whether a business objective is viable and which project must be initiated to implement a business objective (Harvard Business Review, 2010; Robertson, 2004). It is possible that more than one project will be provided as a solution by means of the business case. A business case will be constructed, and various techniques can be used such as net present value, internal rate of return, and payback analysis, as well as benefits realisation. Nonfinancial techniques include weighted scoring and categorisation schemes. The measurement criteria and targets might indicate two potential projects that will achieve the same result. These techniques can be used to determine which project best contributes to the strategies and objectives of the organisation. The end result is a collection of projects that will best achieve the set targets for the strategies and objectives. Another aspect that must be addressed during this step is the elimination of duplicate projects and is addressed by the principles of programme management. The different projects initiated by the business objectives must be grouped together to eliminate duplication.

Once the projects are defined and listed as legitimate projects, they will be managed according to standards and best practices. These standards and best practices are discussed in Chapter 6.

3.2.6 Projects

The previous step determined and defined the projects that should be initiated to realise the business objectives. The success of a specific project will be determined by the measurement criteria, as defined by the business case and the stakeholders.

The next step in the process is to group the projects together into logical groups called programmes.

3.2.7 Programme Management

The determination of programmes is crucial to the organisation because these programmes will be linked to the business objectives determined by the strategy map. Different projects are grouped together to form a programme. These programmes are based upon the internal processes and competence perspectives of the strategy map. This enables the organisation to link projects directly to the strategies. This means that all the projects that form part of the operations management process are grouped together into an operations management programme. Duplicate projects need to be removed, and complementary projects integrated. Duplicate projects are a result of the different measurement criteria. It might happen that the same project was identified to satisfy two different measurement criteria. The main purpose of programme management is to identify and consolidate these duplicate projects in such a manner that the organisation benefits from executing these projects in unity rather than separately. This is an important part of the process, as there are limited resources available to execute these projects. The resulting projects are then formally initiated using the traditional project management processes. The grouping of projects into programmes has the benefit that the programme manager can relate the outcome of the projects directly to the outcome and success of the programme. The net result of the projects will determine the success or failure of the specific programme.

Once the programmes have been defined, it is essential that they be managed in a professional manner. This implies that programme managers must adhere to programme standards and best practices and that they are competent in managing programmes as well.

3.2.8 Programmes

The success of each programme depends on how much it realised the business objectives and benefits linked to it. There will be a direct correlation between the success of the programme and the success with which the business objectives were implemented. The next step in the process is to manage the programmes in such a manner that optimal return is gained by the implementation of these programmes.

3.2.9 Portfolio Management

Portfolio management is about managing the portfolio in such a way that the organisational strategies are optimally implemented and realised (Lin & Hsieh, 2004). The portfolio manager manages the portfolio and makes decisions on the following:

- ***The priority of projects and programmes:*** It might be more beneficial to the organisation to implement one programme before another. The portfolio manager must make decisions on the priority of the programmes.
- ***The inclusion or exclusion of projects and programmes:*** The vision and strategies of the organisation will change over time. It is the responsibility of the portfolio manager to determine which projects and programmes are still necessary to fulfil the strategies and vision. This is also applicable where projects are outsourced to other organisations, such as information technology (IT). IT projects are sometimes outsourced to other organisations that have the knowledge and capabilities to manage a certain IT component. It should be included in the portfolio, as the failure or success of outsourced projects will have an impact on the overall success of the portfolio.

3.2.10 Portfolios

The different projects and programmes form the basis of the portfolio of the organisation. By translating the organisational strategy into projects or programmes, the strategy map as well as the balanced scorecard provides the portfolio manager with a certain number of projects and programmes to include in the portfolio. The portfolio focuses on the financial and customer perspectives of the strategy map. These two perspectives provide the organisational strategies, and the portfolio must ensure that these strategies are implemented in

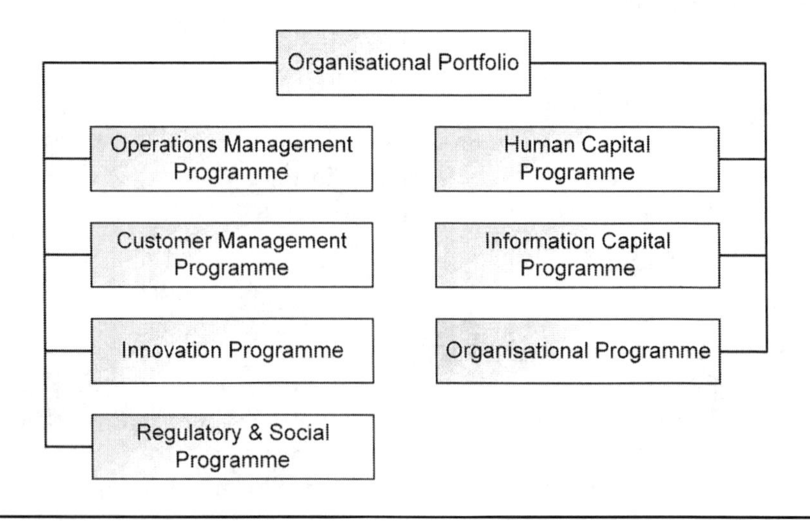

Figure 3-2 Typical Portfolio

such a way that they can be perceived as successful, and, therefore, the vision of the organisation will be realised through the successful implementation of the strategies. The portfolio on the financial perspective of the strategy map must not be confused with the financial portfolio of the organisation. The portfolio and the financial portfolio of an organisation will together realise the financial strategies of the organisation. A typical portfolio is illustrated in Figure 3-2. It is clear from this diagram that a portfolio can consist of up to seven programmes.

The different programmes are managed by a coalition of all the executives within the organisation. The executives will ensure that the different projects within the programme are managed, and that the different programmes are a success.

3.2.11 From Projects Back to Vision

A top-to-bottom process is proposed to turn the vision into projects. On the other hand, a bottom-to-top process is proposed to monitor the achievement of the vision by measuring project performance and progress. The organisation must monitor the progress toward achieving the vision and strategies on a regular basis to ensure that the selected projects produce the expected results. Any deviations need to be addressed by taking corrective action. Monitoring takes place at three levels:

1. The measurement criteria and targets enable the organisation to monitor individual projects. It is the project manager's duty to ensure that the projects achieve these targets.
2. The achievement of business objectives is monitored through the implementation of programmes, as every programme is linked to a business objective. The programme manager must ensure that the programmes achieve these objectives.
3. The successful deployment of strategies depends on the execution of the portfolio. If the portfolio is managed successfully, the organisational strategies will be achieved. The portfolio manager must ensure that this is achieved.

If all the organisational strategies are achieved, then the vision of the organisation is realised. Figure 3-3 illustrates the relationship between and dependency of the vision and the strategies. The vision determines the strategies, and the strategies determine the success of the vision.

The vision takes a long-term view of the organisation, while projects take a shorter-term view. This process enables the organisation to focus on the shorter

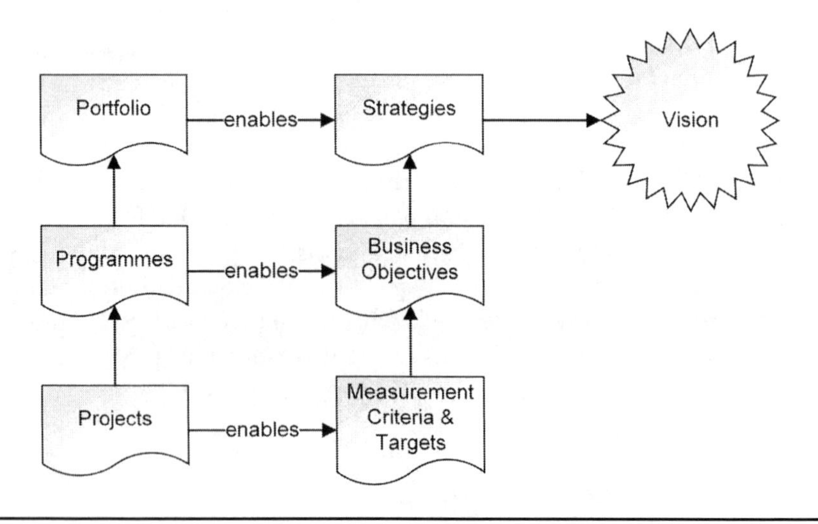

Figure 3-3 Interrelationship between Vision and Strategies

term—that is, the projects—while the success of the short-term projects results in the success of the long-term vision and strategies. Based on the Vision-to-Project Framework, it can be deduced that there is a direct link between projects and the organisational vision and strategies.

3.3 The Executive's Role in the Strategic Alignment of Programmes and Projects

Executives play an important role in the definition of strategies as well as the implementation and realisation of these strategies. The literature has shown that programmes and projects can be used as vehicles to implement strategies. To ensure that strategies become a reality and not just some vague dream, executives must focus on and address the following.

3.3.1 Communicate Vision and Strategies Continuously

It would be ideal if the vision and subsequent strategies would stay static, with as little change as possible. This is, however, not likely to happen as organisational changes occur. Some of these changes include merging with and acquiring other organisations and a change of executives. These changes will have an impact on the strategies, as current strategies might become obsolete, and new strategies are introduced. Whatever the case, when there is a shift or change in the strategies, it is the duty and obligation of the executives to communicate it to the organisation at large and to the programme and project managers specifically. In addition, they need to ensure that their respective programmes and projects are aligned to the new vision and strategy of the organisation. The V2P framework is a tool that can be used to continuously determine which programmes and projects are aligned with the strategies. This framework is useless when the vision and strategies change without informing the people who are actually mandated to implement the strategies that the strategies have changed.

3.3.2 Allocation of Resources

The literature has shown that there is a positive correlation between the successful implementation of strategies and the allocation of resources to programmes and projects. The more resources are allocated to a programme or project, the better the chances are that the programme or project will be a success. A successful project implies the successful realisation of a strategy. It must be noted that resources are not just financial or human. Yes, these two types of resources play a crucial part in the success of a programme or project, and executives should use their political influence to allocate enough of these resources to a programme or project. It is also important that an executive plays an active role during the life cycle of a programme and project as a sponsor. Being a sponsor specifically implies that the executive must provide the programme or project manager with adequate resources. The executive should be involved with the programme or projects that he or she is sponsoring and provide the moral support that is required.

3.3.3 Be Wary of "Pet" Projects

The V2P framework provides a structured mechanism for deriving projects and programmes from the strategies. This should, in turn, prohibit the sponsoring of any so-called pet projects by an executive. "Pet" projects is a phenomenon that pops up every now and then when an executive is trying to push a certain political agenda to score some personal points. This leads, in the long run, to the demise of the organisation and eventually to the individual executive. Executives do get attached to certain projects, as these projects are important for the delivery of strategies for which they are accountable. This affinity must not get to the point at which logic and reasoning are replaced with personal agendas.

3.3.4 Creating Metrics

Metrics are important and should be used by executives to determine the performance of the strategies. Realistic and achievable metrics should be derived and determined for each of the strategies and subsequent business objectives. These metrics should then be used to guide programme and project managers and also to ultimately determine the performance of programmes and projects against the strategies and business objectives. Without any metrics, executives are running blind and will have no idea or sense about how the organisation is performing with regard to strategy implementation.

3.3.5 Business Case

Each project that is initiated based on strategies and business objectives should have a business case associated with it. The purpose of the business case is to set out the rationale for, and justify, any investment in a programme or project. As such, it summarises anticipated benefits while considering alternative options and recommending a preferred solution. It provides a summary of the most important project aspects, such as scope, cost, the time frame, and risks. The business case is owned by an executive who acts as the sponsor. It is the duty of an executive to ensure that valid business cases are produced for each project. These business cases should be linked to the strategies and business objectives and should stipulate how the associated project will realise the strategies and business objectives.

3.4 Conclusion

Evidence provided in this chapter highlights the important role that programmes and projects play in the realisation of organisational strategies. It is almost impossible to realise strategies without programmes and projects. The chapter has provided scientific evidence of the value of programmes and projects when it comes to the implementation of organisational strategies. Literature has also indicated that there is not much research on how this should be done, and the introduction of the V2P framework tries to resolve this issue. The V2P framework can be used to derive programmes and projects from the vision and strategies. This chapter has further highlighted the important role that executives play during the formulation of the vision and strategies as well as in determining which programmes and projects should be used to implement these strategies. Executives might not be directly involved in the day-to-day running of projects, but in their role as sponsors, they are ultimately accountable for the realisation of the strategies.

Executives must realise that, although projects are the vehicles to implement strategy, they are responsible and accountable to clear the road for programmes and projects. This is done through the creation of a proper vision and strategies as well as subsequent metrics that can be used to determine progress. Executives also clear the road to ensure that enough and appropriate resources are available to the respective project managers. Executives cannot determine the vision and strategies and then sit back and relax. They should continue to be involved in the implementation of the strategies as well.

The next chapter focuses on portfolio management and how portfolio management should be implemented within the organisation. Portfolio management ensures that the optimal mix of programmes and projects are implemented for the optimum return on investment.

3.5 References

Bogler, R., & Nir, A. E. (2001). Organizational Vision: The Other Side of the Coin. *Journal of Leadership Studies, 8*(2), 135–144.

Cooke-Davies, T. J., Crawford, L. H., & Lechler, T. G. (2009). Project Management Systems: Moving Project Management from an Operational to a Strategic Discipline. *Project Management Journal, 40*(1), 110–123.

Crawford, L. (2014). Balancing Strategy and Delivery: The Executive View. *Procedia—Social and Behavioral Sciences. 119*(0), 857–866.

Desai, A. B. (2000). Does Strategic Planning Create Value? The Stock Market's Belief. *Management Decision, 38*(10), 685–693.

Dew, N., Goldfarb, B., & Sarasvathy, S. (2006). Optimal Inertia: When Organizations Should Fail. In J. A. C. Baum, S. D. Dobrev, & A. Van Witteloostuijn (Eds.), *Ecology and Strategy* (Advances in Strategic Management), Volume 23 (pp. 73–99). Bingley, UK: Emerald Group Publishing Limited.

Eweje, J., Turner, J. R., & Müller, R. (2012). Maximizing Strategic Value from Megaprojects: The Influence of Information-Feed on Decision-Making by the Project Manager. *International Journal of Project Management, 30*(6), 639–651.

Ezingeard, J. N., McFadzean, E., & Birchall, D. (2007). Mastering the Art of Corroboration: A Conceptual Analysis of Information Assurance and Corporate Strategy Alignment. *Journal of Enterprise Information Management, 20*(1), 96–118.

Harvard Business Review. (2010). *Developing a Business Case.* Boston, MA, USA: Harvard Business Press.

Hermano, V., & Martín-Cruz, N. (2016). The Role of Top Management Involvement in Firms Performing Projects: A Dynamic Capabilities Approach. *Journal of Business Research, 69*(9), 3447–3458.

Jones, G. R. (2001). *Organizational Theory: Text and Cases.* Upper Saddle River, NJ, USA: Prentice Hall.

Joshi, M. P., Kathuria, R., & Porth, S. J. (2003). Alignment of Strategic Priorities and Performance: An Integration of Operations and Strategic Management Perspectives. *Journal of Operations Management, 21*(3), 353–369.

Kakabadse, A., Ward, K., Korac-Kakabadse, N., & Bowman, C. (2001). Role and Contribution of Non-Executive Directors. *Corporate Governance: The International Journal of Business in Society, 1*(1), 4–8.

Kaplan, R. S., & Norton, D. P. (1996). *The Balanced Scorecard.* Boston, MA, USA: Harvard Business School Press.

Kaplan, R. S., & Norton, D. P. (2004). *Strategy Maps: Converting Intangible Assets into Tangible Outcomes.* Boston, MA, USA: Harvard Business School Press.

Lin, C., & Hsieh, P.-J. (2004). A Fuzzy Decision Support System for Strategic Portfolio Management. *Decision Support Systems, 38*(3), 383–398.

Longman, A., & Mullins, J. (2004). Project Management: Key Tool for Implementing Strategy. *Journal of Business Strategy, 25*(5), 54–60.

Manas, J. (2006). *Napoleon on Project Management.* Nashville, TN: Thomas Nelson Inc.

Marnewick, C., & Labuschagne, L. (2006). *A Structured Approach to Derive Projects from the Organisational Vision.* Paper presented at the Project Management Research Conference 2006, Montreal, Quebec, Canada.

Marr, B., & Adams, C. (2004). The Balanced Scorecard and Intangible Assets: Similar Ideas, Unaligned Concepts. *Measuring Business Excellence, 8*(3), 18–27.

Martinsuo, M., Gemünden, H. G., & Huemann, M. (2012). Toward Strategic Value from Projects. *International Journal of Project Management, 30*(6), 637–638.

Martinsuo, M., & Killen, C. P. (2014). Value Management in Project Portfolios: Identifying and Assessing Strategic Value. *Project Management Journal, 45*(5), 56–70.

Nadler, D. A. (2004). What's the Board's Role in Strategy Development?: Engaging the Board in Corporate Strategy. *Strategy & Leadership, 32*(5), 25–33.

Ohmae, K. (2001). *The Invisible Continent: Four Strategic Imperatives of the New Economy.* New York, USA: HarperCollins.

Pearce, J. A., & Robinson, R. B. (2000). *Strategic Management: Formulation, Implementation, and Control.* Boston, MA, USA: McGraw-Hill Higher Education.

Peterson, M. (2002). *Why Are We Doing This Project?* Retrieved from Detroit: https://www.pwc.com/th/en/publications/assets/whydoing.pdf

Project Management Institute. (2009). *Organizational Project Management Maturity Model* (OPM3®) (2 ed.). Newtown Square, PA, USA: Project Management Institute.

Rasiel, E., & Friga, P. D. N. (2001). *The McKinsey Mind.* New York, USA: McGraw-Hill Education.

Richardson, B. (1994). Comprehensive Approach to Strategic Management: Leading Across the Strategic Management Domain. *Management Decision, 32*(8), 27–41.

Ritter, M. (2003). The Use of Balanced Scorecards in the Strategic Management of Corporate Communication. *Corporate Communications: An International Journal, 8*(1), 44–59.

Robertson, S. (2004). Requirements and the Business Case. *IEEE Software, 21*(5), 93–95.

Rowley, J. (2002). Synergy and Strategy in E-Business. *Marketing Intelligence & Planning, 20*(4), 215–222.

Testa, M. R. (1999). Satisfaction with Organizational Vision, Job Satisfaction and Service Efforts: An Empirical Investigation. *Leadership & Organization Development Journal, 20*(3), 154–161.

Thackray, J. (1995). What's New in Financial Strategy? *Planning Review, 23*(3), 14–18.

Van Der Merwe, A. P. (2002). Project Management and Business Development: Integrating Strategy, Structure, Processes, and Projects. *International Journal of Project Management, 20*(5), 401–411.

Young, R., & Grant, J. (2015). Is Strategy Implemented by Projects? Disturbing Evidence in the State of NSW. *International Journal of Project Management, 33*(1), 15–28.

Young, R., Young, M., Jordan, E., & O'Connor, P. (2012). Is Strategy Being Implemented Through Projects? Contrary Evidence from a Leader in New Public Management. *International Journal of Project Management, 30*(8), 887–900.

Chapter 4

Portfolio Management

- Portfolio construction is not a science, more an art and involves lots of judgement. -

— Neil Woodford*

Portfolio management, also sometimes referred to as project portfolio management, is the discipline that organisations use to ensure that all programmes and projects are strategically aligned. The ultimate goal of portfolio management is to select and prioritise programmes and projects to optimise the performance of the portfolio. This is done within the context of strategic alignment. There is a positive relationship between portfolio performance and organisational performance (Müller, Martinsuo, & Blomquist, 2008). The previous chapter highlighted the relationship between a portfolio and the organisational strategies. The success of a portfolio is dependent on the success of the various components that constitute the portfolio. The causal relationship between portfolio and organisational performance emphasises the fact that organisational performance is dependent on portfolio performance. Emphasis should therefore be placed on the success of the portfolio because it will ensure the success of the organisation at the end of the day.

4.1 Defining Portfolio Management

Portfolio management (PfM) is traditionally defined as the management of one or more portfolios in such a way that the organisational strategies and objectives are achieved (Project Management Institute, 2013). The previous chapter

* Retrieved from http://mastersinvest.com/portfoliomanagmentquotes/

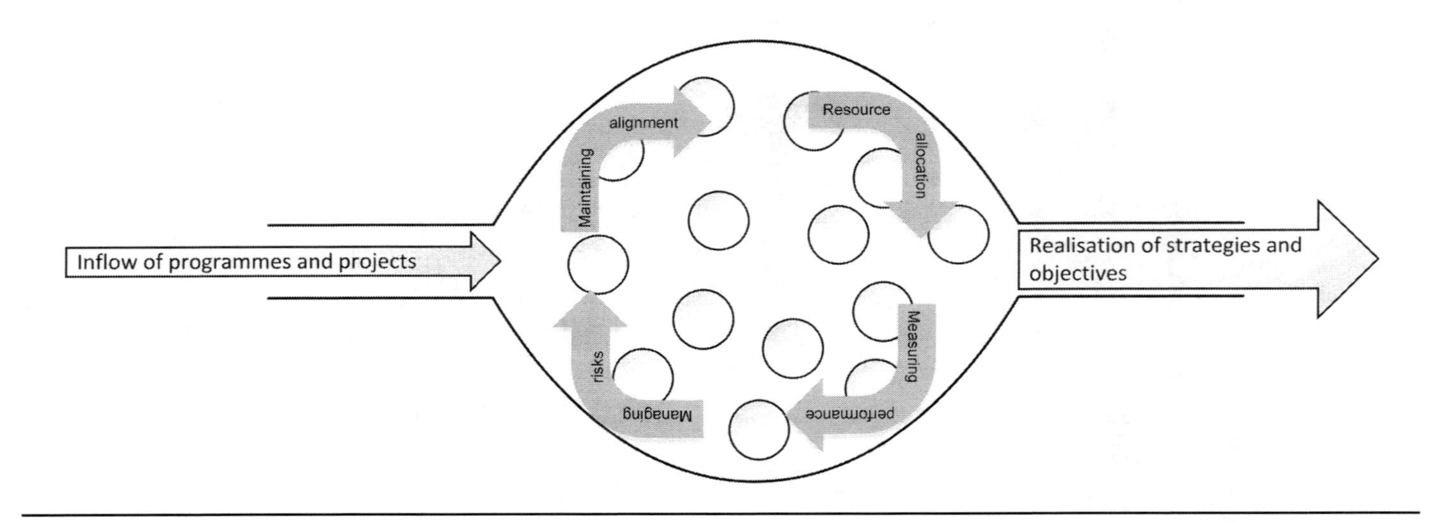

Figure 4-1 Portfolio Management

highlighted the importance of deriving projects from and linking projects with the strategies and objectives. These projects are then grouped together into a portfolio. A portfolio is therefore defined as a collection of programmes and/or projects to achieve the strategic objectives. The realisation of the strategies and objectives is achieved through the performance of the following activities:

- ***Maintaining portfolio alignment:*** The ultimate goal of portfolio management is to continuously maintain alignment between the various programmes and projects and the strategies and objectives.
- ***Allocation of resources:*** Resources are scarce throughout any organisation, and the proper allocation and management of resources should be paramount. Resources include, among others, financial, human, and equipment resources. These resources should be wisely allocated to the various programmes and projects within a portfolio. The allocation should be done in such a way and manner that portfolio alignment is not compromised.
- ***Measuring portfolio performance:*** Throughout the management of a portfolio, the performance should be continuously measured. Performance is measured against the extent that programmes and projects are achieving the strategies and objectives of the organisation.
- ***Managing risks:*** Each programme and project is associated with risks. The culmination of all these risks into a single portfolio of risks might expose the organisation to too many risks. This risk exposure might negatively influence the achievement of the strategies and objectives. Portfolio management therefore focuses on managing risks to minimise risk exposure. The opposite is actually also true where the focus might be on maximising the opportunities, especially when risks are perceived as opportunities.

These four activities must be performed by the organisation on a continual basis to ensure an optimal portfolio, whose sole purpose is to deliver the organisational strategies and objectives. Figure 4-1 provides an overview of the portfolio management activities and the relationship among these activities.

Within the day-to-day management of a portfolio, the portfolio manager is the person who must oversee the process of optimally managing the portfolio. The portfolio manager must focus on doing the right projects and programmes (PMI, 2013).

4.2 Overview of Portfolio Management Standards

This section provides the executive with an overview of the portfolio management standards that are available to organisations. The Project Management

Institute (PMI) defines a standard as a "document, established by consensus and approved by a recognised body, which provides for common and repeated use, rules, guidelines or characteristics for activities or their results, aimed at the achievement of the optimum degree of order in a given context."* A standard can therefore be used by organisations as a means to provide consistency in the way that portfolio management is practiced. The most common standard for portfolio management is that of the PMI.

4.2.1 The Standard for Portfolio Management (PMI, 2013)

The Standard for Portfolio Management (PMI, 2013) describes three process groups and five knowledge areas, as depicted in Table 4-1.

Table 4-1 Process Groups and Knowledge Areas

	Process Groups		
Knowledge areas	**Defining**	**Aligning**	**Authorising and Controlling**
Portfolio strategic management			
Portfolio governance management			
Portfolio performance management			
Portfolio communication management			
Portfolio risk management			

Note: Data derived from *The Standard for Portfolio Management* (PMI, 2013).

A process is a set of actions and activities that are performed to achieve the strategic goal. Each process has inputs, tools, and techniques that are applied and some resulting outputs (PMI, 2013, p. 28). The first process group—*Defining*—focuses on those processes that need to be performed to implement the organisational strategy and objectives. It also consists of processes that determine the structure and roadmap of the portfolio. The *Aligning* process group consists of the processes that need to be executed to manage and optimise the portfolio. Some of the processes include the categorisation, evaluation, selection, and

* http://www.pmi.org/pmbok-guide-and-standards/standards-overview.aspx

management of the portfolio components. The last process group, *Authorising and Controlling*, consists of the processes needed to authorise the portfolio and provide continuous oversight.

The five knowledge areas can be summarised as follows:

- ***Portfolio strategic management:*** Each portfolio should have a strategic plan and a roadmap describing how the organisational strategies and objectives should be implemented. This knowledge area assesses and manages the continuous alignment of the strategic plan and roadmap to the strategies and objectives (PMI, 2013).
- ***Portfolio governance management:*** The processes within this knowledge area focus on the overall oversight of the portfolio. Specific processes include the definition, optimisation, and authorisation of the portfolio. Chapter 10 deals with project governance, and portfolio governance management should be performed in support of the overall governance of projects (PMI, 2013).
- ***Portfolio performance management:*** The focus of this knowledge area is to plan, measure, and monitor the value of the portfolio. This is done through the realisation of the organisational strategies and objectives. The objective is to determine the optimal composition necessary to best achieve these strategies and objectives (PMI, 2013).
- ***Portfolio communication management:*** This knowledge area includes the processes that are needed to develop the communication management plan as well as manage the portfolio information (PMI, 2013, p. 105).
- ***Portfolio risk management:*** Portfolio risks are assessed and analysed with the goal to capitalise on potential opportunities, on the one hand, and to mitigate those events or activities that can adversely impact the portfolio, on the other hand (PMI, 2013). The objective of portfolio risk management is to accept the right amount of risk for the optimum delivery of the portfolio.

4.2.2 ISO 21504:2015 Project, Programme, and Portfolio Management—Guidance on Portfolio Management

ISO 21504:2015 provides guidance on the principles of project and programme portfolio management (International Organization for Standardization, 2015). ISO 21504:2015 is relevant to any type of organisation—public or private—and any size organisation or sector. ISO 21504:2015 is not as comprehensive as *The Standard for Portfolio Management* (PMI, 2013) and consists of three major sections.

1. The first section focuses on principles of portfolio management. The first principle covers the notion of the context and need for portfolio management. The rationale for the adoption of portfolio management is provided by this principle. The second principle provides an overview of portfolio management and briefly covers aspects such as portfolio management itself, the portfolio structure, capabilities, and constraints, as well as opportunities. The third principle focuses on the various roles and responsibilities. The fourth principle covers stakeholder engagement.
2. The second section highlights the prerequisites for portfolio management. The justification for portfolio management is provided, as well as a framework that defines how the organisation is going to determine and decide which components will form part of the portfolio and which ones will be removed. This section also looks at the various types of components and how these components should be categorised, evaluated, selected, and prioritised. The alignment of the portfolio with the various organisational processes and systems is covered, and various processes are mentioned. The establishment of a management system should provide visibility and relevant information to the decision makers. The eight prerequisites address the notion of the performance-reporting structure. The purpose of the structure is to enable the monitoring and tracking of both the performance as well as the achievement of business strategies and objectives. The last two prerequisites address the notion of improving portfolio management itself and the governance of the portfolio.
3. The third section actually addresses how portfolios should be managed and briefly addresses the following processes:
   - Setting portfolio objectives
   - Identifying potential portfolio components
   - Defining the portfolio plan
   - Assessing and selecting portfolio components
   - Validating portfolio alignment to strategic objectives
   - Evaluating and reporting portfolio performance
   - Balancing and optimising the portfolio

There are some similarities between ISO 21504:2015 and *The Standard for Portfolio Management* (PMI, 2013), but the PMI® standard is much more detailed.

4.2.3 Management of Portfolios (MoP®)

The *Management of Portfolios* (MoP®) was first published in 2011, by the United Kingdom government (Jenner & Kilford, 2011). This standard describes two portfolio cycles, as illustrated in Figure 4-2.

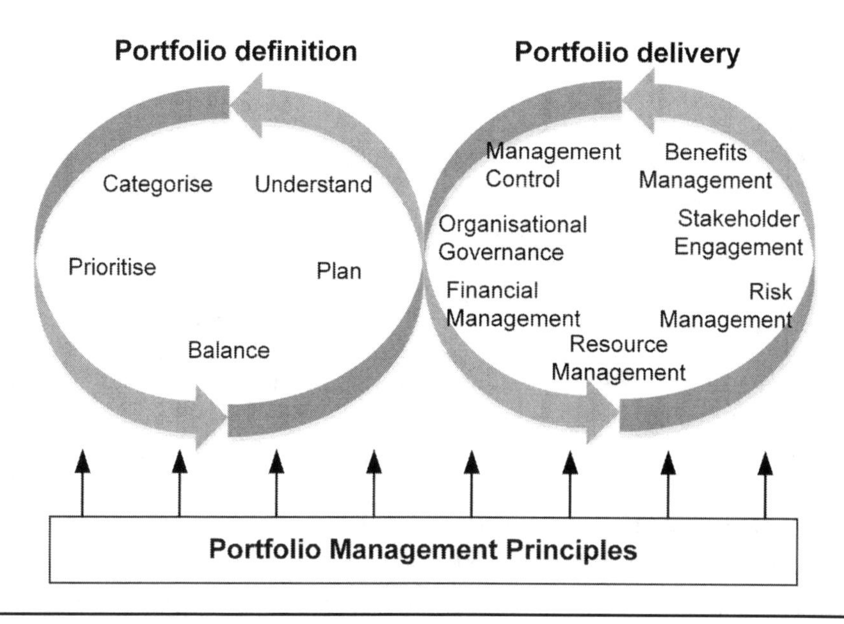

Figure 4-2 Management of Portfolios (MoP®)

The two cycles comprise practices that are found within two continuous portfolio management cycles.

1. The ***Portfolio Definition Cycle*** focuses on doing the right things. Information is collated to provide clarity to senior management and the wider audience with regard to the collection of change initiatives and how these initiatives will deliver the greatest contribution to the strategic objectives. The first practice, ***Understand***, provides the recognition of portfolio components that already exist or that need to exist. The ***Categorise*** practice organises the components into groups based on the strategies and business objectives. Next, ***Prioritise*** ranks the components based on one or more agreed-upon measures. The ***Balance*** practice ensures that the portfolio is balanced in terms of contributing to the strategies and business objectives, the business impact, as well as risk exposure. The last practice is ***Plan***. Information is collated from the portfolio definition cycle to create a portfolio strategy, which culminates in the second cycle—the portfolio delivery plan, which should be agreed upon by the organisational leaders.
2. The ***Portfolio Delivery Cycle*** focuses on "doing those things right." Its purpose is the successful implementation of the planned components as agreed upon during the portfolio definition cycle. The seven practices within the portfolio delivery cycle include ***Management Control***, wherein individual and portfolio-level decisions are made regarding the

progress against the portfolio delivery strategy and plan. The ***Benefits Management*** practice identifies and manages the benefits that should be realised from the portfolio as well as the contribution of each of the portfolio components to operational performance, strategies, and business objectives. ***Financial Management*** ensures that portfolio processes and decisions are aligned to the financial management cycle, and financial considerations form a key element in all decisions. ***Risk Management*** ensures the consistent and effective management of the portfolio's exposure to risk at both the individual and collective level. The ***Stakeholder Engagement*** practice ensures that the needs of the portfolio's customers are identified and managed appropriately. ***Organisational Governance*** ensures that portfolio management governance is aligned with the wider organisational governance structure, enabling clear understanding of all decisions. ***Resource Management*** puts mechanisms in place to understand and manage the amount of resources required to deliver the changes.

Five portfolio management principles are the foundation of the two portfolio cycles. These principles provide the organisational environment in which the portfolio definition and delivery practices can operate effectively (Jenner & Kilford, 2011; Kilford, 2016).

1. ***Senior Management Commitment:*** The executive management of an organisation should publicly champion and positively communicate the value of portfolio management while actively participating in the processes.
2. ***Governance Alignment:*** Without proper governance, portfolio management will fail. This principle focuses on the structures for escalation and decision making. The PfM structure should integrate with the existing organisational decision-making processes.
3. ***Strategy Alignment:*** The portfolio must contain components that contribute toward the achievement of the organisational strategies and objectives. MoP® suggests a driver-based model starting with a high-level strategy, down to strategic objectives, then benefits, and, finally, change initiatives that will deliver them. This is in line with the V2P framework described in Chapter 3.
4. ***Portfolio Office:*** A portfolio office function should exist, with a reporting line directly to the executive management. The purpose of the portfolio office is to provide standards, analysis, and enhanced collaborative working. The portfolio office should be able to provide up-to-date and accurate information to allow effective decision making by portfolio managers.

5. ***Energised Change Culture:*** Executive management should create a culture in which the people are motivated and striving to do things as best as is possible, they believe in the organisation's goals, and they feel like part of the organisation. An environment for open communication that includes listening and engagement with staff should be established.

It is for the organisation to decide which of these standards it wants to follow. This decision must be made within the context of the organisation as well as that of the country.

4.3 Portfolio Management Process

The process required to manage a portfolio is fairly standard and is illustrated in Figure 4-3. The entire process is governed by the organisation's current vision

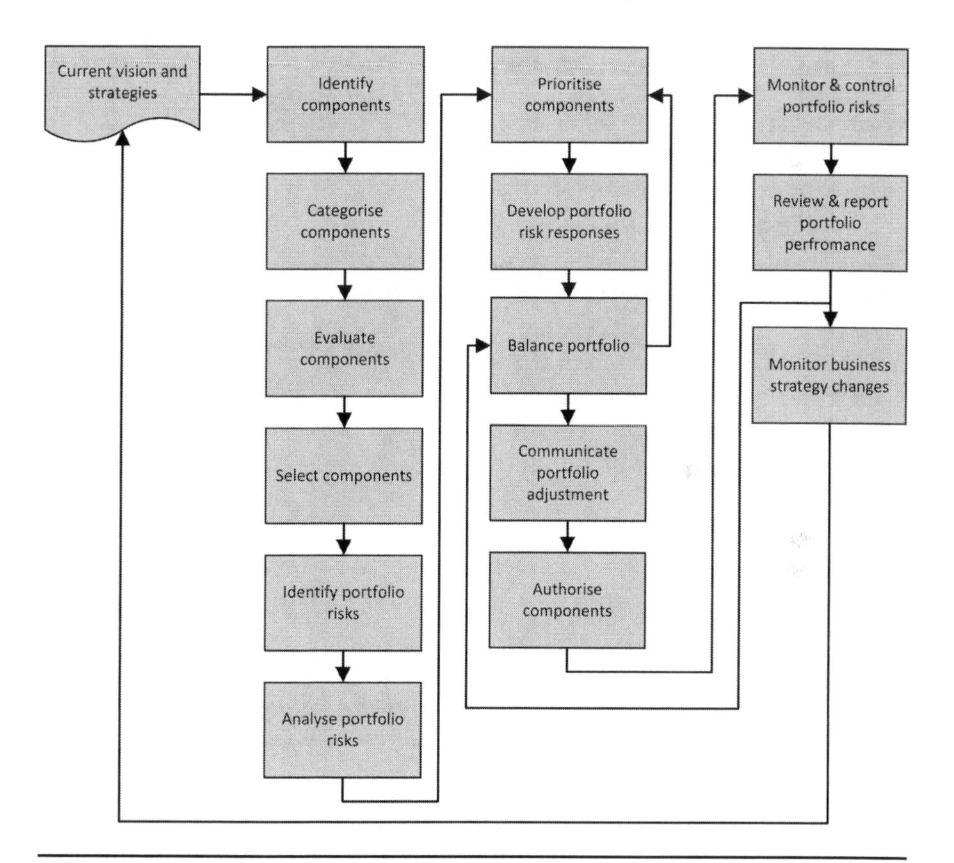

Figure 4-3 Portfolio Management Process

and strategies. This provides direction for the portfolio, and the portfolio manager uses it to balance the portfolio. The first step is to identify all the components that, when successfully implemented, will contribute to the successful realisation of the strategies. During the second step, the components can be categorised into different categories. Categorisation might be as per Figure 3-2, where the categories are based upon the balanced scorecard. Component categorisation assists in balancing the portfolio because the portfolio manager can verify whether one category is overpopulated while another category might be underpopulated. It also assists with risk analyses because one does not want to overexpose a portfolio to one specific category. The third step is to evaluate each and every component to determine whether it should be incorporated into the portfolio.

According to the PMI (2013), the portfolio manager applies a series of evaluation criteria. It must be noted that these criteria must be quantifiable in order to measure, rank, and prioritise each and every component. Some evaluation criteria that can be used include financial and nonfinancial benefits, regulatory compliance, as well as the capabilities and capacities of human resources. It must be noted that nonquantifiable criteria, wherever possible, must be translated into quantifiable criteria. This assists the portfolio manager in making decisions in an unbiased way and manner. Once the components are evaluated, the next logical step is to select the components that will form part of the portfolio.

The selection of the components will depend on various constraints such as the availability of financial and human resources. A positive correlation between the selection of components and portfolio performance does exist. The identification of portfolio risks comprises two parts. The first part is to determine the overall risk associated with the portfolio itself. The second part determines the risk associated with each individual component. The next step is to evaluate the risks of each component and create a comprehensive risk portfolio. The purpose of this risk portfolio is to determine the risk exposure of the portfolio at large. Portfolio risk management manages all the components' risks in an aggregated pool, adding and removing components. This is done to maintain a predetermined and acceptable level of risk tolerance and exposure (Maynard, 2015). Analysing the portfolio in terms of its risk tolerance and exposure provides insight into which components should eventually form part of the portfolio. The idea is to manage the risk tolerance and exposure in such a way and manner that the portfolio does not compromise the future of the organisation.

Given the steps of identifying, evaluating, selecting, and determining the risk exposure, the next step is to prioritise the components in such a way that the portfolio is optimised. The prioritisation must be done to ensure the realisation of the vision and strategies. The prioritisation is the prerogative of the portfolio manager, but the process of prioritising components must be transparent as well as consistent. Studies found that the prioritisation of components is a key success factor (Beringer, Jonas, & Kock, 2013; Costantino, Di Gravio, & Nonino, 2015).

Once the components are prioritised, a risk response must be developed for the entire portfolio. The risk response is for the portfolio and not for the individual components that now constitute the portfolio. Risk management should and must still be done for each of the individual components. The portfolio is then balanced based on the priorities of the components and the risk responses and in respect to diverse strategies of the organisation. These strategies include the financial, organisational, and operational strategies, and the portfolio should be balanced in such a way and manner that each of these strategies is realised. Once the portfolio is balanced, the portfolio must be communicated to the organisation at large. According to Turner and Lecoeuvre (2015), the purpose of this communication is to win the support of the stakeholders. The portfolio manager needs to sell the benefits and persuade the stakeholders to commit themselves and resources to the portfolio itself as well as to the individual components. This speaks directly to the communication competence of the portfolio manager. The portfolio components are then authorised and implemented based on the prioritisation of the components within the portfolio.

The last part of the portfolio management process focuses on the day-to-day management of the portfolio. During this part, the portfolio is continuously evaluated against the portfolio's risk exposure and the performance of each of the individual components. The portfolio needs to be reviewed regularly, and if there is any deviance, then the portfolio must be re-evaluated. This re-evaluation might include the reorganisation of the components, which includes the prioritisation and balancing of the components. The portfolio also needs to be re-evaluated if there are any changes to the vision and strategies of the organisation.

4.4 Portfolio Management Success

Meskendahl (2010) is of the opinion that the successful management of a portfolio needs to contribute to the realisation of the organisational strategies. The success of the portfolio will be measured based on four basic criteria, as follows: the (i) the maximization of the financial value of the portfolio; (ii) the linking of the portfolio to the organisational strategies; (iii) balancing the components within the portfolio; and (iv) the success of each individual component. To determine whether a portfolio is managed successfully, the results must be measureable and have to cover a comprehensive array of measurements (Martinsuo & Lehtonen, 2007; Meskendahl, 2010). This view is shared by

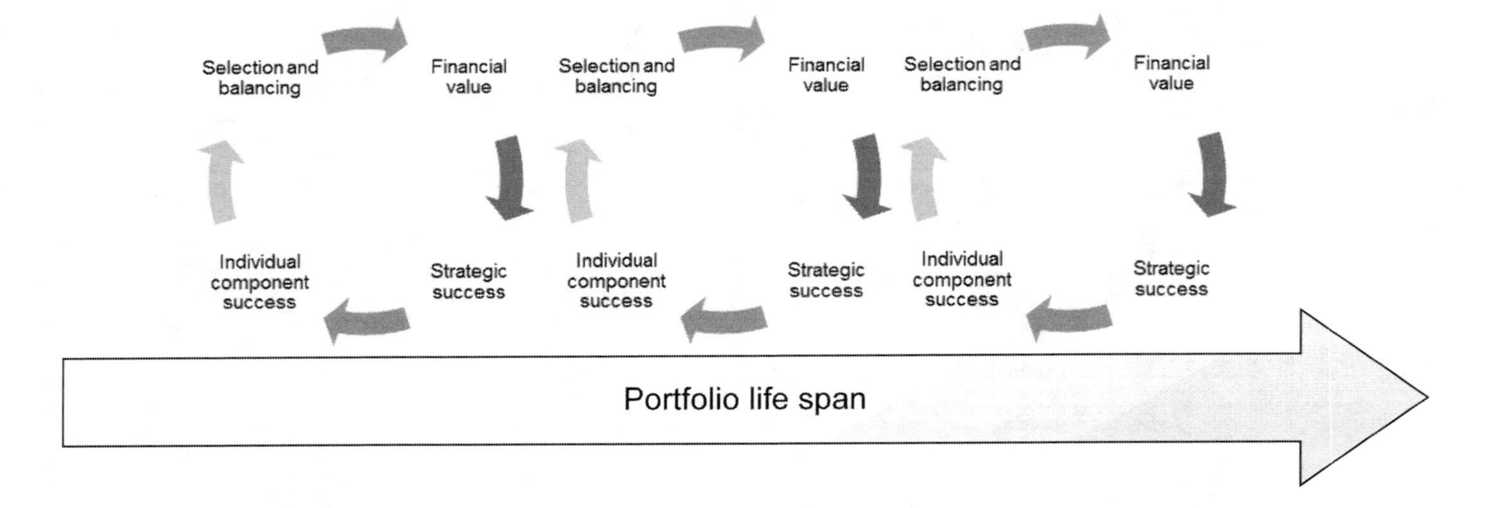

Figure 4-4 Portfolio Management Success

Beringer et al. (2013), who stipulate that the success of the portfolio is equal to the individual component's success and the strategic fit of the portfolio. The success of the portfolio can be divided into four separate success criteria, as per Figure 4-4.

To successfully achieve the first criterion, the portfolio must succeed in maximising the financial value of the organisation. Second, the portfolio must contribute to the overall strategic success of the organisation. Third, the portfolio will only be successful if the individual components in it are successful. The fourth criterion focuses on the selection of the components and the balancing of the portfolio. This was discussed in the previous section. The success of the portfolio is measured over an extended period of time, which might cover a couple of years. The idea is that a portfolio is perceived as a permanent organisation and should therefore be treated as such. Marnewick (2014) provides an extensive model for evaluating the success of a portfolio as well as the factors that contribute to portfolio success.

4.5 The Role of the Portfolio Manager

The role of the portfolio manager evolves around three main responsibilities that comprise the portfolio management process. These three responsibilities are the establishment, continuous monitoring, and management of the portfolio. A more detailed breakdown of a portfolio manager's responsibilities include the following (De Reyck et al., 2005):

1. Defining the goals and objectives of the portfolio is important so that everyone understands what the portfolio is expected to achieve. This responsibility touches on governance because it provides guidance and steering for everyone involved in the portfolio.
2. Another important role entails the coordination of the portfolio, which includes participation in various steering committees and other relevant meetings (Blomquist & Müller, 2006). Resource management falls within this role and focuses first on resource identification and second on the procurement of these identified resources. The continuous evaluation of portfolio components constitutes a part of portfolio coordination. Programmes and projects that do not contribute to the overall value of the portfolio should be terminated.
3. Managing a portfolio involves various expectations and requirements from the stakeholders. One of the most important responsibilities of a portfolio manager is to understand, accept, and make trade-offs to the overall benefit of the portfolio and ultimately that of the organisation.

4. As per Figure 4-1, risk management plays an important role in a portfolio. It is the responsibility of the portfolio manager to identify, eliminate, minimise, and mitigate risks associated with the portfolio.
5. The ultimate goal of a portfolio is to deliver on the strategies of the organisations. This can only be achieved if the performance of the portfolio is continuously monitored. This implies that the progress of the portfolio's components must be determined in relation to the progress made toward the achievement of the goals and objectives.
6. The most important role and responsibility of a portfolio manager is to establish confidence. The team as well as all the stakeholders must have confidence in the portfolio manager. This confidence is grounded in the fact that the portfolio manager can achieve the desired objective of the portfolio.
7. Portfolio managers should also perform relatively mundane activities that are not necessarily directly linked to the management of the portfolio but which eventually contribute to its success. Blomquist and Müller (2006) identified some of these activities—for example, performing administrative tasks and handling issues, which culminates in improving efficiency and the coaching of programme and project managers.

The portfolio manager is ultimately accountable for the optimal performance of the portfolio. To perform this role to the best of his or her ability, the portfolio manager should display certain competencies and skills. The *Individual Competence Baseline for Project, Programme & Portfolio Management* (IPMA ICB®) of the International Project Management Association (IPMA®) provides guidelines on the competencies that portfolio managers should possess (IPMA, 2015). The competencies are divided into three major categories—Perspective, People, and Practice.

The summary of the portfolio management competencies that follows is per the IPMA ICB. The first grouping of competencies focus on the how the portfolio manager perceives the portfolio within the context of the organisation. The portfolio manager has to have perspective competencies that address the context of the portfolio. The first competence (**strategy**) focuses on the portfolio manager's ability to understand and transform the strategies into manageable elements using projects. The **governance competence** determines whether the portfolio manager understands and aligns the portfolio with established structures, systems, and processes within the organisation. This alignment provides support for portfolios and influences the way they are managed. The **compliance, standards, and regulations** element describes how a portfolio manager interprets and balances external and internal restrictions imposed on the portfolio—for example, a country, company, or industry. Portfolio managers need to understand the role of **power and interest** so that he or she recognises and understands the difference and relationship between personal and group interests as well as the resulting politics and use of power. The last competency within the perspective competence group addresses **culture and values.** The emphasis is on the portfolio manager's approach to influencing the organisation's culture and values and, ultimately, the wider society in which the portfolio is situated. The rationale is to influence the organisation to have a more positive attitude toward portfolio management.

The competence area "people" deals with the personal and social competencies of the individual portfolio manager. The first competency, **self-reflection,** is the ability to reflect on the portfolio manager's emotions, behaviour, and values and to determine his or her impact on the portfolio at large. A portfolio manager must also exhibit **personal integrity and reliability** because a lack of these qualities may lead to portfolio failure. **Personal communication** addresses the ability to acquire, analyse, and disseminate information in a transparent and consistent way.. Portfolio managers liaise with various stakeholders on a daily basis and should therefore be competent in **relations and engagement**: The portfolio manager should portray **leadership** by providing direction and guidance to the team at large. This leadership competency also involves the ability to choose and apply appropriate styles of management in different situations. The portfolio team consists of various stakeholders who are working together to realise the strategies of the organisation. The portfolio manager should be competent in **teamwork** because he or she needs to bring the team together to realise the portfolio objectives. **Conflict and crisis** occur during the management of a portfolio. The important question should be how the portfolio manager solves conflicts and crises by being observant and presenting remedies for disagreements. **Resourcefulness** is another competency within the armoury of a portfolio manager. Managing a portfolio on a day-to-day basis comes with its own challenges and problems. Portfolio managers should have the ability to apply various techniques and ways of thinking to address these challenges and problems. As stated earlier, there are various stakeholders within a portfolio, and these stakeholders might have opposing opinions or requirements. Portfolio managers make use of **negotiation** to balance these different opinions, requirements, and expectations. The above-mentioned people competencies have one purpose and that is to achieve results. Through the application of all these competencies, a degree of **results orientation** is achieved in order to obtain the optimum outcome for all the parties involved.

The third competence area is "practice" and covers the core competencies of a portfolio manager. The portfolio manager should be able to master 14 competencies within this area. The first competency focuses on how a **portfolio is designed**. The design takes the requirements of the stakeholders into consideration as well as the strategies of the organisation and translate them into a high-level portfolio design to ensure the highest probability of success. The second practice competency focus on the **benefits** that should be achieved when the portfolio is managed in a successful way. Not all programmes and projects will be incorporated into a portfolio, and that is where the **scope** competency plays a role. The portfolio manager should define the specific scope of the portfolio as well as which components are not to be included in the portfolio. The exclusion of certain programmes and projects is dependent on the overall balancing of the portfolio, as illustrated in Figure 4-3. Although a portfolio and the management thereof, is not necessarily **time dependent**, all the components should be structured in such a way as to optimise the execution of the individual components. Within the domain of portfolio, programme, and project management, programmes and projects are perceived as temporary organisations, whereas a portfolio is perceived as a permanent organisation. One of a portfolio manager's competences is the ability to focus on the **organisation and information sharing** of and between the programmes and projects within the portfolio. **Quality** is an important competence that a portfolio manager should portray. First, quality is portrayed in the way that the portfolio was organised, and, second, quality is portrayed in the outputs of the various components. Competence in **finance** includes all activities required for managing the financial resources of the portfolio and all the components within the portfolio. A separate competence element is that of specific **resources,** which includes defining, acquiring, controlling, and developing the resources that are necessary to realise the portfolio's outcomes. Portfolio **procurement** is a process of buying or obtaining goods and/or services from external parties. It includes all the relevant processes from planning and making the necessary purchases up to the administration and closure of contracts. Once all the components are selected and prioritised, they must be implemented. A balanced **plan** is needed to implement the components in a controlled manner. Managing a portfolio does not come without **risk** from the portfolio itself as well as from the various components. The portfolio manager must constantly analyse the risk exposure of the portfolio, the impact on the organisation itself, and the realisation of the strategies. The **stakeholders'** competence element includes identifying, analysing, engaging and managing the attitudes and expectations of all relevant stakeholders. Programmes and projects institute change by nature. By implication, a portfolio should also bring about change to the organisation. This change will be in the form of newly developed capabilities, with their associated benefits. **Change** and **transformation** provide the portfolio manager with the processes, tools, and techniques that should be utilised to assist organisations in making a successful transition, resulting in the adoption and realisation of change. **Selecting and balancing** is the last competence within the practice competence area. This competence focuses on the assessment, selection, and performance monitoring of the various components within the portfolio, as well as on the balancing of the portfolio to ensure that the portfolio creates continuous optimal benefit for the organisation.

4.6 The Executive's Role in Portfolio Management

The role of the executive with regard to portfolio management is increasingly important, and executives should be aware of and knowledgeable about the discipline of portfolio management and what it entails. The rationale is that portfolio management ensures a balanced view of what needs to happen to implement the strategies and business objectives. For that reason, executives need to either be involved in the management of the portfolio or put the following enablers in place to ensure the successful management of the portfolio.

4.6.1 Standardisation of Portfolio Management

Portfolio management and the role of the portfolio manager must be standardised within the organisation. The benefit of such a standardisation is that everyone will speak the same language, and there will be no confusion in terms of processes and terminology. The organisation at large must decide which portfolio management standard is the one that best fits its strategies. This task is best left for the portfolio manager to decide, but it is the responsibility of the executive to ensure that this standard is continuously adhered to and enforced when there is a deviance from the adopted processes. Deviations to the adopted standard will lead to inconsistency, which in turn will lead to internal organisational conflict because people may feel that they have been treated differently.

4.6.2 Competent Portfolio Managers

Section 4.5 highlighted the competencies of a portfolio manager. The role of a portfolio manager is a senior role within an organisation, and in most instances, he or she might report directly to an executive. Portfolio managers should therefore be competent in managing a portfolio. Executives' role is twofold. First, executives should encourage and also demand that portfolio managers are certified against the standard that the organisation is following. Although this certification itself does not make a portfolio manager competent, it provides the executive with the guarantee that the portfolio manager is (i) knowledgeable about portfolio management; and (ii) the portfolio manager belongs to a community of practitioners. Second, it is the duty of executives to ensure that the portfolio managers are continuously evaluated based on their competencies. Where gaps are identified, remedial action should occur to enhance or improve the portfolio manager's competencies.

4.6.3 Communication of Organisational Strategies

It was mentioned in Section 3.3.1 that executives must communicate the vision and strategies of the organisation. This is becoming more important within portfolio management because the portfolio is directly influencing and being influenced by the vision and strategies. Executives have an obligation to continuously ensure that the current strategies are communicated and that the portfolio adheres to the strategies. Where deviations are encountered, corrective action must be taken, but ultimately, executives are accountable for the implementation of the strategies.

4.6.4 Corporate Politics

Implementing the Project Portfolio: A Vital C-Suite Focus (PMI, 2015) reveals that corporate politics plays an important role within portfolio management, especially during the selection process of components. It was found that portfolio components were selected based on substantial negotiations between powerful executives and other senior members, instead of following a process. This study also reveals that executives follow their own interests, and "pet" projects ultimately undermine formal portfolio management. Executives should be aware of corporate politics and the negative role it plays during the management of a portfolio. It therefore becomes unnegotiable that a formal portfolio management process be implemented and adhered to. Deviations to the formal processes should not be tolerated, and all executives should adhere these processes, regardless of their position.

4.6.5 Information Management

To defuse the impact of corporate politics, reliable information must be presented at all times to reflect the status of the portfolio as well as that of the individual components. To accomplish this, key performance indicators (KPIs) for the portfolio should be established. They can be based on the success criteria for a portfolio, as highlighted in Section 4.4. Executives should determine and agree upon these KPIs. It is the responsibility of the portfolio manager to put metrics in place to measure these KPIs and report on them. Portfolio management information should improve the effectiveness of the portfolio and minimise political influences because all decisions will be based on relevant and accurate information.

The report from the PMI (2015) provides more insight into the responsibilities of executives with regard to portfolio management and emphasises that executives cannot be aloof but should be actively involved in portfolio management.

4.7 Conclusion

This chapter highlighted the importance of portfolio management and the important role that it plays in realising the organisational vision and strategies. Portfolio management is becoming the de facto process to implement strategies and select the appropriate programmes and projects to fulfil this role. The chapter began with a high-level view of portfolio management and what it constitutes. In general, portfolio management consists of a few basic processes that ensure the optimal portfolio that will ultimately deliver on the strategies of the organisation. Various portfolio management standards were also discussed to indicate that organisations have a choice as to which standard to use. The final decision might be based on geographical location, but once a standard is selected, it is advisable for the organisation to adhere to the processes and not change them often. Based on these standards, a generic portfolio management process is presented. This enables executives to understand the essence of portfolio management without going into the details of it. The next section discusses the success of a portfolio and how this success can be measured and determined. It is important to continuously measure the success of the portfolio because the success of the organisation is directly linked to the success of the portfolio. Portfolio managers play a vital role in portfolio management, and the competencies are briefly discussed to indicate the wide range of competencies that a portfolio manager must master. Portfolio management cannot be successful without the support of executives, and the chapter is concluded by giving executives some guidance on the positive role they can play within portfolio management.

4.8 References

Beringer, C., Jonas, D., & Kock, A. (2013). Behavior of Internal Stakeholders in Project Portfolio Management and Its Impact on Success. *International Journal of Project Management, 31*(6), 830–846.

Blomquist, T., & Müller, R. (2006). Practices, Roles, and Responsibilities of Middle Managers in Program and Portfolio Management. *Project Management Journal, 37*(1), 52–66.

Costantino, F., Di Gravio, G., & Nonino, F. (2015). Project Selection in Project Portfolio Management: An Artificial Neural Network Model Based on Critical Success Factors. *International Journal of Project Management, 33*(8), 1744–1754.

De Reyck, B., Grushka-Cockayne, Y., Lockett, M., Calderini, S. R., Moura, M., & Sloper, A. (2005). The Impact of Project Portfolio Management on Information Technology Projects. *International Journal of Project Management, 23*(7), 524–537. http://dx.doi.org/10.1016/j.ijproman.2005.02.003

International Organization for Standardization. (2015). *ISO 21504:2015 Project, Programme, and Portfolio Management—Guidance on Portfolio Management* (p. 13). Geneva, Switzerland: International Organization for Standardization.

International Project Management Association. (2015). *Individual Competence Baseline for Project, Programme & Portfolio Management*, Version 4. Zurich, Switzerland: International Project Management Association.

Jenner, S., & Kilford, C. (2011). *Management of Portfolios*. London, UK: Stationery Office.

Kilford, C. (2016). *MoP® and PMI®'s Portfolio Management Guidance: Is This the Yin Yang of Portfolio Management?* Axelos: London, United Kingdom. Retrieved from https://www.axelos.com/CMSPages/GetFile.aspx?guid=ef9997b5-9235-4080-8bde-c0e5cba37032

Marnewick, C. (2014). Portfolio Management Success. In G. Levin & J. Wyzalek (Eds.), *Portfolio Management: A Strategic Approach* (pp. 123–148). Boca Raton, FL, USA: CRC Press.

Martinsuo, M., & Lehtonen, P. (2007). Program and Its Initiation in Practice: Development Program Initiation in a Public Consortium. *International Journal of Project Management, 25*(4), 337–345.

Maynard, D. A. (2015). The Accuracy of Portfolio Risk Management. In G. Levin & J. Wyzalek (Eds.), *Portfolio Management: A Strategic Approach* (pp. 165–174). Boca Raton, FL, USA: CRC Press.

Meskendahl, S. (2010). The Influence of Business Strategy on Project Portfolio Management and Its Success—A Conceptual Framework. *International Journal of Project Management, 28*(8), 807–817.

Müller, R., Martinsuo, M., & Blomquist, T. (2008). Project Portfolio Control and Portfolio Management Performance in Different Contexts. *Project Management Journal, 39*(3), 28–42.

Project Management Institute. (2013). *The Standard for Portfolio Management* (3 ed.). Newtown Square, PA, USA: Project Management Institute.

Project Management Institute. (2015). *Implementing the Project Portfolio: A Vital C-Suite Focus.* Newton Square, PA, USA: Project Management Institute. Retrieved from http://www.pmi.org/learning/thought-leadership/series/portfolio-management/project-portfolio-c-suite-focus

Turner, R., & Lecoeuvre, L. (2015). Marketing the Project Portfolio. In G. Levin & J. Wyzalek (Eds.), *Portfolio Management: A Strategic Approach* (pp. 175–192). Boca Raton, FL, USA: CRC Press.

Chapter 5

Programme Management

~ The whole is greater than the sum of its parts. ~

— Aristotle*

The second level of management within the P3 is programme management. Programme management is an interesting concept because it is not always necessary to have a programme. When it is possible to achieve the planned and/or perceived outcomes with a project, then it is advisable that organisations initiate a project rather than a programme. Programmes, and for that matter programme management, are more complex to manage than projects, and it is sometimes better to reduce the complexity and have various individual projects.

Sometimes, however, the best results can only be achieved through a programme, and it is therefore necessary for executives to understand the importance of programme management and the role that it plays within the P3 discipline. This chapter provides the information that is necessary regarding programme management, including the definition of programme management, the processes of programme management, the role of the programme manager, and how executives can measure the success of a programme.

The first section covers the concept and management of programmes.

5.1 Defining Programme Management

There is a common understanding that a programme consists of two or more projects that are logically grouped together. This grouping cannot be made

* Retrieved from http://philosiblog.com/2016/03/17/the-whole-is-more-than-the-sum-of-its-parts/

arbitrarily, and the aim should be for these grouped projects to deliver benefits to the organisations that would not be possible if these projects were managed individually. In simple terms, one plus one should be three or higher. If one plus one is still two, then these projects should be managed as individual projects because it would not make logical sense to group them together. Unlike a portfolio that can consist of various programmes, projects, and other work that is normally considered operational activities, a programme consists only of projects and, in rare instances, perhaps operational activities as well. A key reason for grouping projects into a programme is to generate benefits for the organisation. Benefits and the management thereof are briefly discussed later on in this chapter, and are discussed in detail in Chapter 8.

A distinction must be made between a large project that might consist of smaller projects and a programme that consists of various related projects. Pellegrinelli, Partington, Hemingway, Mohdzain, and Shah (2007) argue that when an organisation views programmes as large projects, it loses most of the benefits sought in setting up programmes in the first place. Thiry (2010) claims that project and programme management can be distinguished based on ambiguity and uncertainty. Large or complex projects are subject to high uncertainty and should be managed according to project management principles. On the other hand, when there is high level of ambiguity because projects are not well defined and have a high level of complexity, these initiatives are actually classified as programmes and should be managed according to programme management principles (Thiry, 2010). A clear differentiation is provided by Heaslip (2014) who states that programmes pursue value through strategies that rely on uncertain outcomes, whereas large projects deliver value via activities that have predictable outputs and/or outcomes.

Programme management is defined as the "application of knowledge, skills, tools, and techniques to a programme to meet the programme's requirements and to obtain benefits and control not available by managing projects individually" (Project Management Institute [PMI], 2013, p. 6). Programme management then involves the alignment of multiple projects to achieve the business objectives, and it also allows for optimal cost, scheduling, and effort. Artto, Martinsuo, Gemünden, and Murtoaro (2009) conclude that the PMI's definition of programme management is as good as any other because it states the basics—that is, projects are grouped together to achieve benefits. From the various definitions of programme management, it can be deduced that programme management entails the following:

- Programme management is about the management of projects and not about project management.
- Projects within a programme are formally related to each other. This formal relation between the projects implies that the successful completion

of all projects accomplishes more than the successful completion of each individual project.

- Programme management focuses on realising benefits.
- Programme management achieves a set of business objectives, as discussed in Chapter 3.

5.1.1 The Rationale for Programme Management

Why should an organisation, and, for that matter, executives, invest in programmes and programme management? The rationale for programme management can be summarised in the following points:

- ***Responsibility:*** A programme manager is the only person responsible for the entire initiative, thus leaving the engineering, marketing, and other functional managers free to focus on their specific deliverables. No other person can be held responsible if a programme is a failure. This eliminates the shift of blame because one person is in charge of a programme.
- ***Tight linkage with the business objectives of the organisation:*** A programme manager has the responsibility to meet the business objectives. The programme manager ensures that the entire programme management team stays focused and delivers profitable outcomes that are consistent with market, customer, and portfolio requirements.
- ***Speed:*** Programme managers are committed to optimising the different but related projects within the programme. They streamline the different projects by adjusting the progress of the different projects. This is achieved by transferring resources from projects on track to projects in need. The programme manager achieves a balance among the different projects by allocating resources where they are needed within projects. At the end of the day, this balance should ensure that the programme itself will be a success. With balancing, resources are not wasted because they are shared and moved among the projects to achieve the optimum resource utilisation.
- ***Objectivity to ensure cross-functional integration:*** Programme managers provide a "neutral corner" that shifts the dialog in the organisation from "What is best for my project?" to "What is best for the programme?"
- ***Integration of value:*** Delivering value to the customer is key to the success of any new product and/or service. Only the programme manager can ensure that every deliverable is understood and is contributing to the value proposition. Most importantly, the programme manager always makes sure that the customer is represented.
- ***Repeatability:*** Programme managers have the charter and skills to refine the programme life cycle and ensure repeatability. Programme managers

have an eagle-eye view of all the projects. This allows them to determine shortfalls within a project and ultimately within the programme. The programme manager can ensure that all the shortfalls are eliminated in future programmes of the same nature. This will lead to the optimisation of the programme and the refinement of the life cycle.

- ***Team morale:*** Programme managers should be astute readers of the mood of the project managers who are part of the programme. A good programme manager can anticipate problems, resolve conflicts, and maintain the enthusiasm of the team of project managers. This is done even during the toughest phases of the programme.
- ***Managing the extended team:*** Programmes might involve participants from within the organisation, strategic partners, suppliers, consultants, and contractors. A programme manager should have collaboration skills and the ability to focus on effectively coordinating the various contributors to the programme.

The above illustrates that there is a need for the introduction of a programme management level within the organisation. Programme management provides the organisation with a vehicle to manage and monitor the implementation of business objectives. Without programme management, an organisation will not know if a business objective was implemented and whether the benefits were realised.

5.2 Programme Management Framework

The programme management framework consists of two sections. The first section contains the key areas of programme management. The focus of the key areas is to determine the programme scope, the aim and purpose of the programme, and the approach the programme manager must follow to ensure a successful programme implementation. The second section of the framework defines the processes involved to ensure that the programme stays on course and within the set parameters of the programme. If a group of related projects are not managed as a programme, then it is likely that it will run off course and fail to achieve the desired outcome. The combination of the key areas and the processes forms the programme management framework.

5.2.1 Key Areas

According to Haughey (2001), there are eight key areas that guide a programme, and together with the programme management processes, they provide the programme manager with a framework to manage programmes.

- ***Vision:*** The vision is the high-level strategy that drives an organisation toward a goal or desired outcome.
- ***Aims and Objectives:*** This is a more detailed statement that explains exactly what is required. It provides a point of reference to go back to when renewed focus is required.
- ***Scope:*** The scope provides boundaries to the programme and explains exactly what it is that will be delivered.
- ***Design:*** This is the way that projects are put together to make up a programme. The programme manager decides which projects are dependent upon each other, which projects can run concurrently, and which must run last.
- ***Approach:*** This is the way the programme will be run. This is entirely up to the programme manager.
- ***Resourcing:*** This looks at the allocation and usage of resources across the programme as well as within each specific project.
- ***Responsibility:*** Each area of the programme must be identified, and responsibilities must be allocated to these areas. Every member of the programme team must clearly understand his or her responsibilities and those of his or her team members.
- ***Benefits Realisation:*** This is the process at the end of the programme by which the benefits identified at the beginning of the programme are measured.

5.2.2 Processes

Programme management can also be divided into processes, as illustrated in Figure 5-1. These processes assist the organisation and the programme manager, in particular, to focus on particular aspects of the programme.

The processes of programme management are as follows:

- ***Programme Identification:*** Chapter 3 explained the relationship between business objectives and programme management. A programme is identified to implement a specific business objective. The business objective is derived from the organisational strategy that, in turn, was derived from the vision of the organisation. The identification of a programme is a typically

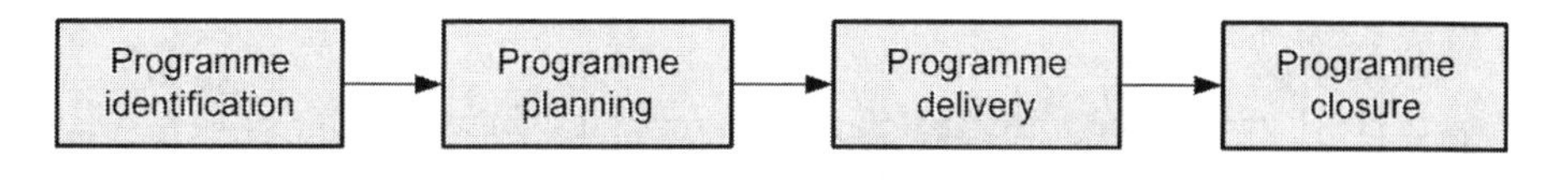

Figure 5-1 Programme Management Processes

short process whose purpose is to define the overall scope of the programme. It initiates the development of initial thoughts on the vision, anticipated benefits, risk and estimated costs, time scales, and effort required, which help to determine whether the programme should proceed. This also includes establishing a programme mandate, which provides the high-level requirements of the programme and the benefits required by the organisation. An initial vision is refined that is a description of the transformed organisation as a result of delivering this programme of change. Activities during this phase provide the detailed information that establishes the definition of the new capabilities, the way they are going to be delivered, details of how the programme will be run, changes implemented, and benefits delivered. The programme is thereby clearly defined, and the organisation decides whether to formally commit to the programme or not.

- ***Programme Planning:*** If approval to proceed has been obtained, this process is used to appoint individuals to programme management and support roles and to set procedures and infrastructures in place to support the programme. The programme manager will
  - *Define clear objectives:* All the objectives of the programme must relate to the implementation of a business objective. The definition of the objectives will allow the programme manager to distinguish between business objectives that will realise the organisational strategy and business objectives that will not.
  - *Agree on roles and responsibilities with the team:* The team consists of different project managers. Each project manager will be in charge of his or her own project, and it is important to select a project manager to fit a specific project. Some project managers are more able to implement change, whereas other project managers are more able to implement physical information technology infrastructure.
  - *Agree on priorities of the projects that make up the programme:* All the projects within the programme need to be successful for the programme to be successful. It is also true that some projects are of higher priority than other projects. This is the case where one project is dependent on the outcome of another project.
- ***Programme Delivery:*** Once the programme is established, this process will provide an effective monitoring and management governance structure for the different projects within the programme such that the required outputs from the projects are delivered. This process covers the prime areas of interaction with project management and requires active coordination of the programme's components, the relationship with other programmes, and the interface with the business strategy. This also means that programme documentation needs to be updated and revisited regularly to reflect any changes. The programme activities, work streams, and projects

will deliver new capabilities, services, or business operations. The purpose of this process is to track the specific outcomes that were identified at the start and to drive through the process of achieving these outcomes within the business operations during the life of the programme. Throughout the programme, progress is reviewed at defined milestones, and the benefit profiles and benefits plan are adjusted to reflect any changes. This process also manages the transition between old and new ways of working. The responsibilities for the achievement of outcomes must be clearly assigned. The programme manager needs some method of measuring the overall status of the programme and will make use of digital dashboards. These digital dashboards provide a high-level view of the status of the programme as well as the status of the individual projects within the programme.

- ***Closing a Programme:*** In order to establish and maintain a clear focus on achieving the business objectives, a programme should have a predetermined endpoint and a formal process of closedown. This includes confirmation that the changes defined have been delivered and the business strategy achieved. Benefits management may need to continue beyond the formal closure of the programme, particularly if the achievement of measurable improvements is to be part of ongoing business operations. It must be noted that the life span of the programme exceeds the life span of the projects within the programme. This means that although the projects are completed successfully, some time must elapse before the programme can be classified as a success or a failure.

It is clear that there are actually two timelines within a programme. The first timeline is each individual project's timeline and is relatively short in comparison with the programme timeline. This clearly illustrates that the programme continues long after the individual projects are completed. The second timeline is that of the programme, which spans all the individual projects' timelines and beyond to the point where benefits are realised. The key areas are also allocated to each process. The key areas—vision, aims, and scope—form part of the programme identification process. These key areas enable the programme manager to identify the programme and document the scope within which the programme will function, as well as the business strategy that must be fulfilled at the end of the programme. The programme planning process uses the key areas: design, approach, resourcing, and responsibilities. This phase will provide the blueprint for the programme, and all programme evaluation will be made against the blueprint derived from the programme planning. The third process is the physical delivery of the programme. This is achieved by the implementation of all the projects that are part of the programme. The programme manager can allocate resources from one project to another project if he or she sees that one of the projects is behind schedule. The programme manager must balance all of the projects in order for the sum total of the projects to be a success. One project within the programme might be a failure in the sense that it did not realise all of the intended benefits, but the success of the programme is not measured on just one project but on the sum total of all the projects. The last phase within the programme management framework is the closure of the programme. This phase occurs after all of the projects are completed and the outcome of the programme is measured against the blueprint. The purpose of a programme is to implement business objectives and benefits, and the provided framework enables a programme manager to do this effectively.

5.3 Overview of Programme Management Standards

There are currently two programme management standards available to organisations—*The Standard for Program Management* (PMI, 2013) and *Managing Successful Programmes* (Sowden, 2011). The International Organization for Standardization (ISO®) is currently working on a programme management standard, but it is not yet released for public use.

5.3.1 The Standard for Program Management (PMI, 2013)

This standard consists of performance domains that need to be adhered to when organisations are implementing programmes. A performance domain is "complementary groupings of related areas of activity, concern, or function that uniquely characterise and differentiate the activities found in one performance domain from the others within the full scope of program management work" (PMI, 2013, p. 17). The five performance domains are as follows:

1. ***Programme strategy alignment:*** This performance domain identifies all the programme outputs and outcomes to provide benefits that are consistent with the organisation's goals and objectives. Programmes are designed to align with the organisational strategy and to ensure that the promised benefits are realised.
2. ***Programme benefits management:*** This performance domain defines, creates, maximises, and delivers the benefits provided by the programme. This is accomplished through the following processes: (i) benefits Identification; (ii) benefits analysis and planning; (iii) benefits delivery; (iv) benefits transition; and (v) benefits sustainment.
3. ***Programme stakeholder engagement:*** The third performance domain identifies and analyses the needs of the stakeholders and manages

expectations as well as communications to foster stakeholder support. The stakeholder engagement domain proceeds through three activities—(i) stakeholder identification; (ii) stakeholder engagement planning; and (iii) stakeholder engagement.

4. ***Programme governance:*** This performance domain enables and performs programme decision making, establishes practices to support the programme, and maintains programme oversight. Programme governance covers the systems and methods by which a programme and its strategy are defined, authorised, monitored, and supported by its sponsoring organisation.
5. ***Programme life cycle management:*** This performance domain manages the programme activities required to facilitate effective programme definition, program delivery, and program closure. This domain presents the phases of the program life cycle—(i) the programme definition phase; (ii) the programme benefits delivery phase; and (iii) the program closure phase.

5.3.2 Managing Successful Programmes (Sowden, 2011)

The MSP® framework is based on three core concepts:

1. ***Principles:*** These are derived from positive and negative lessons learned from programme experiences. They are the common factors that underpin the success of any transformational change. There are seven principles that form part of this concept:
   - *Remaining aligned with corporate strategy:* Programmes need to stay aligned with the organisational strategies, even if those strategies change over time. Programmes take strategic drivers and use them to govern the respective components and business change activities.
   - *Leading change:* Programme managers act as "agents of change" and lead others through the changes their programme plans to deliver.
   - *Envisioning and communicating a better future:* Programmes should define a clear vision of the desired future state they seek to create as part of the transformational change they will deliver to the organisation. The programme's vision should be defined early in the program life cycle and refined as necessary over time.
   - *Focusing on the benefits and threats to them:* Programmes satisfy business objectives by achieving their desired outcomes and delivering the resulting benefits of the programme to the organisation. These benefits should be relevant to the strategic context of the organisation in order to add value.

- *Add value:* Programmes add value to the organisation by exceeding the benefits provided by the straightforward sum of their constituent components and major activities.
- *Designing and delivering a coherent capability:* Programme capabilities are defined in the programme blueprint and delivered according to the programme plan. Keeping a handle on program scope and quality allows the programme to release its incremental improvements with minimal operational impact.
- *Learning from experience:* Programme teams are expected to learn from previous experience in order to improve their performance over time. Lessons learned are sought, recorded, and acted upon throughout the programme life cycle, which typically happens at major review points.

2. ***Governance Themes:*** Nine themes constitute the governance or control framework that defines an organisation's approach to programme management. These themes allow an organisation to put the right leadership, delivery team, organisational structures, and controls in place, which provides the best chance for success.
   - The ***organisation*** governance theme describes the overall structure as well as the individual roles and responsibilities of the programme. The establishment of an effective programme organisation is critical to the success of the programme.
   - The ***vision*** of the programme describes the future state envisioned when the programme objectives are successfully delivered. The vision statement describes new services and improvements and is used to gain commitment and buy in from the stakeholders. The programme vision must not be confused with the organisational vision.
   - The governance theme relating to ***leadership and stakeholder engagement*** is crucial for a successful programme because without the adequate support from the right people, a programme may lack the resources that it needs to fulfil the promised benefits.
   - The purpose of ***benefits management*** is the clear identification of benefits, and the use of these benefits is a roadmap for the programme. Benefits management is discussed separately and in detail in Chapter 8.
   - The ***blueprint*** is an expansion of the organisational vision and models the organisation of the future. The blueprint describes in detail the working practices and processes, the required information input, as well as any supporting technology.
   - The ***programme plan*** provides in detail how the programme is to be run. It provides information about resources, risk management,

individual projects, deadlines, constraints, and scheduling. Control, on the other hand, ensures that the programme remains on track and delivers the promised benefits, and any accompanying transition is effected smoothly and profitably.

- The programme's ***business case*** intends to answer the question whether the specific programme is worth the required investment? The information aggregated into the business case includes the value of the benefits, the associated risks, and the estimated timescale for achievement.
- ***Risks*** are uncertain events that would influence the outcome of a programme either positively or negatively. An issue, on the other hand, is a risk that has occurred. Programme risk management consists of identifying risks, the assessment of these risks, planning on how to deal with the identified risks, and implementing a plan to minimise risk exposure.
- ***Quality management*** focuses on the efficacy of the programme and its benefits in achieving the strategic goals that are at stake. Effective quality management is essential to ensure that programme benefits are successfully and comprehensively realised.

3. ***Transformational Flow:*** This provides a route through the life cycle of a programme, from its conception through to the delivery of the new capability, outcomes, and benefits. This core concept consists of six iterative and interrelated processes.
   - *Identifying a programme:* This is the organisational vision that triggers the change and the identification of the programme that turns the vision and business objectives into a tangible business concept.
   - *Defining a programme:* This process provides input into the decision as to whether to continue with the programme. Detailed planning occurs during this step.
   - *Managing the tranches:* This process implements the governance strategies and ensures that the capability delivery is aligned to the strategies and that the benefits are realised.
   - *Delivering the capability:* This process focuses on the components and how to coordinate and manage the delivery of each component in relation to the overall programme plan.
   - *Realising the benefits:* This process manages the benefits, from identification through to the successful realisation.
   - *Closing the programme:* This process ensures the formal closing of the programme and occurs when the programme has delivered the required new capabilities, as described in the blueprint.

5.4 The Role of the Programme Manager

The role of the programme manager differs from that of the portfolio manager and the project manager. The role of the programme manager is not a natural evolution from that of a senior project manager. These two types of roles are distinctly different from each other, and promoting proven project managers into the role of programme management can be disastrous (Partington, Pellegrinelli, & Young, 2005). A competent programme manager can positively influence the success of programmes (Miterev, Engwall, & Jerbrant, 2016).

The IPMA ICB® addresses the competencies of the programme manager as it did those of the portfolio manager. There is no distinction in the different competencies that sets a portfolio and programme manager apart. The distinction is in the level of competence that is needed. The most common competencies that a programme manager should display are summarised as follows (International Project Management Association [IPMA], 2015; Levin & Ward, 2011; Miterev et al., 2016):

- ***Leading:*** The programme manager should use a variety of approaches to lead the programme and the respective project managers and also ensure that everyone understands the importance of the programme and what it needs to achieve.
- ***Relationship building:*** The programme manager needs to work actively in building relationships with all the stakeholders. The purpose is to build trust among the stakeholders and the team members.
- ***Negotiating:*** The various stakeholders bring their own preconceived ideas and beliefs to the programme, and these might be in conflict with that of other stakeholders. The programme manager needs to seek solutions that build winning solutions, and this is achieved through negotiation.
- ***Critical thinking:*** This competence involves the use of higher-level, analytical, and abstract, as well as open thinking in order to make better informed decisions.
- ***Facilitating:*** A programme is the ideal environment for conflict, and the programme manager must create an atmosphere among the team members and stakeholders that facilitates a positive environment focusing on delivering a successful programme.
- ***Mentoring:*** The programme manager is leading and managing various team members, including project managers. The programme manager should encourage and facilitate the professional development of each team member.
- ***Embracing change:*** The main purpose of a programme is to bring about change in the organisation. Programme managers must be comfortable with change and be an advocate for change.

- ***Communicating:*** The programme manager should use various communication skills and media to communicate with a wide range of stakeholders. Communication might occur at an executive level but also at a very technical level.
- ***Distilling key information:*** A lot of information is generated during the lifespan of a programme. An important competence is to process the information internally and to deliver the key messages in a clear and concise way to the team members and stakeholders.
- ***Flexibility:*** Programme managers need to be flexible in the way that they approach the various projects within the programme and should portray situational leadership and flexibility.
- ***Decision making:*** Programme managers should exhibit competence in decision making. They should be able to make difficult decisions from the onset of the programme and show confidence in their own decisions.
- ***Managing programmes:*** Programme managers should also be competent in managing the programme from initiation to closure and, ultimately, the realisation of the promised benefits.
- ***Contextual awareness:*** This competence focuses on the way that the programme manager interprets and understands the programme's long-term effects and implications for the organisation as a whole.
- ***Courage:*** As advocates of change, programme managers should portray courage and stand up and defend, or argue for, their programmes and team members.

The various competencies are captured in Figure 5-2 and highlight the complexity of the competencies that programme managers must display.

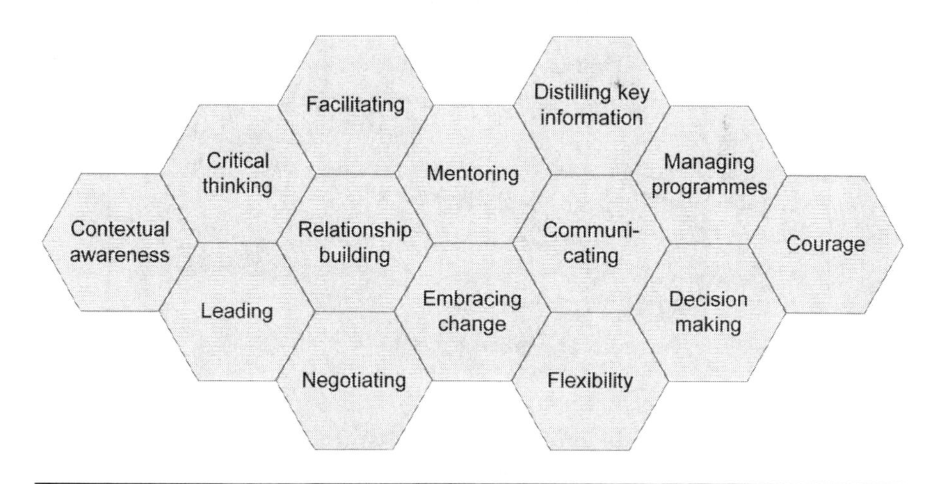

Figure 5-2 Programme Manager Competencies

5.5 Programme Management Success

Literature describing programme management success is scarce. In *The Standard for Program Management*, the PMI (2013) states that the programme governance board must decide on the criteria that should be used to determine when a programme is successful.

The first success criterion focuses on benefits realisation and the maximum delivery of all of the benefits. The implication is that a programme's success can sometimes only be measured after the formal closure of the programme because benefits are realised after official programme closure. The actual realised benefits are measured against the promised or intended benefits. There might be a difference between the promised benefit and the realised benefit, and the level of tolerance should form part of the success criteria.

Unlike a portfolio, whose sole intention is to deliver on the vision and strategies of the organisation, programmes focus on the successful delivery of business objectives. Successful business objectives should therefore be the second success criteria. As stated earlier, in Chapter 3, business objectives focus on changing processes internal to the organisation. It is therefore logical that the success of the programme is also determined based on the change that was achieved. The questions that must be answered are whether a change did occur, was the change sufficient to improve on internal processes, and are these improvements actually benefiting the organisation at large.

5.6 The Executive's Role in Programme Management

The role of the executive is not as evident in programme management as it is within portfolio management. Portfolio management is directly linked to the vision and strategies of the organisation, and therefore executives are more involved with portfolio management. Programme management focuses on strategic objectives and attracts the attention of middle managers but not necessarily that of the executives. Executives, however, still play an important role in some governance-related aspects.

5.6.1 Rationale for a Programme

Programmes are not large projects divided into smaller projects. But programmes are also not just a couple of projects lumped together. The rationale for a programme is that the projects must be grouped together to achieve the effect of one plus one equals three or more. This itself places a responsibility on the organisation to ensure that each programme is actually a programme and meets the criteria of a programme. Executives must create an environment in which programmes can be formed when and as needed. The fact that there are programme managers within an organisation does not always imply that there should be programmes to manage. This creates somewhat of a dilemma, especially within smaller organisations where there might not always be programmes.

5.6.2 Clear Career Path for Programme Managers

Partington et al. (2005) state that programme management is not the next logical step in the promotion ladder. They explain that one cannot be promoted from a senior project manager to a programme manager. The rationale is that programme managers need a different skill set and different competencies from a project manager. This is highlighted in Section 5.4, where these competencies are discussed. Executives must then ensure that there is a career path within the organisation for P3 managers. This career path must cater to two aspects. The first aspect is that each path—portfolio, programme, and project management—must be distinct from the others, and this difference should be evident in the roles and responsibilities of the respective managers. The second aspect that needs to be considered is the career path of a programme manager. Some individuals will not be able to make the switch from project management to programme management or from programme management to portfolio management. They will be caught in a specific career path's most senior level, and the question that needs to be considered is whether this is adequate.

5.6.3 Enable Programme Managers to be Successful

Going hand-in-hand with the well-structured career path is the support that executives provide to programme managers to be successful. This is achieved through training, tools, and resources. Through the provision of these three basic elements, an environment is created whereby programme managers can be successful. Programme team members should be assigned permanently to the programme to ensure that change occurs. Change cannot happen when team members are assigned part-time to the programme. In addition, programme managers must be well versed in programme management and should have the right skills and competencies, as described in Section 5.4. The competencies must be well entrenched. A good starting point is to provide training for programme management certification.

5.6.4 *Measuring the Progress of the Programme*

Unlike a portfolio that has various success criteria, the success of a programme is measured based on the realisation of the promised benefits and the achievement of the business objectives, which lead to change. To achieve this, regular measures should be in place to determine the progress of the programme. PWC (2014) found in a global research report that the tracking of programmes does not take place. This then creates a concern because no one knows the progress of the programme and the subsequent realisation of benefits. To curb this, executives should enforce regular progress measurements and make decisive decisions when everything is not going according to plan.

5.7 Conclusion

This chapter addressed the concept of programme management and where it fits within the organisational structure. First, the notion of programmes and programmes were addressed, and the conclusion is that programmes are different from large projects and have a unique role to play within the organisation. The chapter then discussed a generic way to manage programmes so that organisations can deliver on the intended benefits and business objectives associated with the programme. The various programme management standards were then discussed. This provides executives with an overview of the options available to organisations with regard to standards and potential certifications for programme managers. This flowed into the following section, which focuses on the role that the programme manager fulfils within the organisation and the P3 discipline. It also highlights the competencies that a programme manager must master to become proficient. The chapter concluded with a section on the criteria for measuring programme management success—benefits realisation and the achievement of the business objectives.

The chapter highlighted the importance of programme management within the organisation. Programme management is the glue between portfolio and project management, and organisations do not have the luxury to disregard or negate the discipline of programme management. Executives should become familiar with the programme management discipline and put all the necessary enablers in place to empower programme managers and the discipline at large. Programme management is not necessarily perceived as a mature discipline, regardless of all the standards and certified programme managers. The onus rests on the executives to empower the programme managers.

5.8 References

Artto, K., Martinsuo, M., Gemünden, H. G., & Murtoaro, J. (2009). Foundations of Program Management: A Bibliometric View. *International Journal of Project Management, 27*(1), 1–18.

Haughey, D. (2001). A Perspective on Programme Management. Retrieved from http://nead.co/files/program-management-study.pdf

Heaslip, R. J. (2014). *Managing Complex Projects and Programs: How to Improve Leadership of Complex Initiatives Using a Third-Generation Approach.* Hoboken, NJ, USA: John Wiley & Sons.

International Project Management Association. (2015). *Individual Competence Baseline for Project, Programme & Portfolio Management,* Version 4.0 (pp. 416). Zurich, Switzerland: International Project Management Association.

Levin, G., & Ward, J. L. (2011). *Program Management Complexity—A Competency Model.* Boca Raton, FL, USA: CRC Press.

Miterev, M., Engwall, M., & Jerbrant, A. (2016). Exploring Program Management Competences for Various Program Types. *International Journal of Project Management, 34*(3), 545–557.

Partington, D., Pellegrinelli, S., & Young, M. (2005). Attributes and Levels of Programme Management Competence: An Interpretive Study. *International Journal of Project Management, 23*(2), 87–95.

Pellegrinelli, S., Partington, D., Hemingway, C., Mohdzain, Z., & Shah, M. (2007). The Importance of Context in Programme Management: An Empirical Review of Programme Practices. *International Journal of Project Management, 25*(1), 41:55.

Project Management Institute. (2013). *The Standard for Program Management* (3 ed.). Newtown Square, PA, USA: Project Management Institute.

PWC. (2014). *4th Global Portfolio and Programme Management Survey.* Retrieved from http://www.pwc.com/en_GX/gx/consulting-services/portfolio-programme-management/assets/global-ppm-survey.pdf

Sowden, R. (2011). *Managing Successful Programmes.* London, UK: The Stationery Office.

Thiry, M. (2010). *Program Management.* Surrey, England: Gower Publishing Ltd.

Chapter 6

Project Management

~ Project management is the art of creating the illusion that any outcome is the result of a series of predetermined, deliberate acts when, in fact, it was dumb luck. ~

— Harold Kerzner*

Project management is the third and foundational level of portfolio, programme, and project management. Without project management, and for that matter projects, organisations are doomed for failure. Portfolio and programme management focus on the strategic levels where the aim is to deliver on the organisational strategies and associated business objectives. To realise the strategies and business objectives, projects need to be executed and managed. Strategies and business objectives are eventually realised through the successful management of projects. Projects deliver either a service or product that is utilised by organisational leaders to achieve business objectives and realise the strategies.

Thus, the importance of project management cannot be negated, and executives should realise its importance. Project management plays the role of an enabler within an organisation, and to successfully fulfil this role, organisational leaders should remove all stumbling blocks that might inhibit the successful delivery of projects. This chapter provides guidelines that can be used to ensure the successful management of projects. First, the concepts of projects and project management are discussed to provide a common understanding of what project management entails. Second, various standards and methodologies are discussed. New methodologies are pushing the boundaries of project management, especially in the field of information technology (IT). Projects cannot be implemented without qualified and competent project managers, and

* Retrieved from https://gcimmarrusti.wordpress.com/pm-quotes/

the focus of the fourth section in this chapter is to determine what competencies a project manager should possess. Project management maturity and project success go hand in hand and are discussed in the fifth and sixth sections. This chapter should assist executives in determining the requirements for the successful delivery of projects.

6.1 Defining Project Management

Various definitions exist to explain projects and project management. The definition that is used most widely is the one by the Project Management Institute (PMI), which states that a project is temporary in nature and that a unique product, service, or result is delivered at the end (PMI, 2013). The definition implies first that a project must at one point in time stop, as it is temporary. Second, every product or service created through a project is unique. This distinguishes a project from operations, where the product or service is not unique but is one of a repetitive nature. Through its PRINCE2® methodology, the Office of Government Commerce (OCG) (2009) adds another dimension to the definition of a project by stating that a project is not just temporary, but it is actually a temporary organisation within the permanent organisation.

A project needs to be managed according to some principles, and therefore project management can be defined as the application of knowledge and skills, as well as tools and techniques, to the various project activities or tasks to meet the project requirements (PMI, 2013). First, project managers must possess knowledge and skills to manage a project. They must also be able to apply various tools and techniques (Mnkandla & Marnewick, 2011). Second, various activities are involved in the creation of the project's product or service. Third, and maybe most importantly, is the fact that there must be adherence and conformance to certain predetermined requirements.

It is evident that some form of formality is necessary. Al-Khouri (2015) categorically states that organisations are more successful when project management is used than when it is not used or ignored. He continues by discussing the various benefits of project management, which include, among others, efficiency, customer satisfaction, the creation of a competitive edge, and capacity building. Andersen (2016) believes that a project, as a temporary organisation, should provide value to the permanent organisation and that this should be the ultimate goal of a project.

6.1.1 Project Life Cycle

The project life cycle is a collection of project phases and defines what work is to be performed in each phase (Kloppenborg, 2015; Milosevic & Srivannaboon,

2006; Schwalbe, 2010). The project life cycle normally consists of four to six phases, but the application of these phases differs from organisation to organisation. The most common phases are as follows:

- ***Concept phase:*** This is the first phase in the project life cycle, and it gives a brief description of the project itself (Schwalbe, 2010; Tummala & Burchett, 1999). It includes aspects such as the deliverables of the project and the preliminary costs. This phase focuses on the identification of the operational needs, business drivers, strategic plans, and other factors that define the scope and objectives of the project.
- ***Development phase:*** The aim of this specific phase is to elaborate on the deliverables of the concept or initial phase (Besner & Hobbs, 2006). The project team creates detailed project plans, and more accurate costing and scoping can be done since additional information is available to the project team. This phase provides an in-depth work breakdown structure (WBS), typically up to three levels deep. This allows the project team to make more accurate assumptions about the project.
- ***Implementation phase:*** During this phase, the project team delivers the product, as specified in the previous two phases, and provides performance reports to the stakeholders (Jugdev & Müller, 2005). It is also the phase that normally takes the longest to complete and is the most expensive phase in terms of time and cost (Jugdev & Müller, 2005; PMI, 2013).
- ***Close-out phase:*** Client acceptance plays an important role in the close-out phase (Hyväri, 2006), and all the work is completed during this phase (Schwalbe, 2010). The completed product or service must be related back to the original specifications defined in the concept phase. A post-implementation audit is performed to measure the effectiveness of the project against the goals and objectives.

The project life cycle plays an important role because it provides the project manager as well as the organisation with a structured way of managing projects. It does not matter whether the project life cycle consists of four or six phases, as long as the phases cover the work that must be done throughout the project life cycle.

6.1.2 Process Groups

The process groups are divided into five groups (Marchewka, 2009; PMI, 2013). These five process groups depend on one another and follow the same sequence for every project. The aim of these five process groups is to deliver a specific result, and each of these processes occurs continuously during the project life cycle.

- ***Initiating process:*** The initiating process group focuses on the beginning of a project or life cycle phase. It requires the organisation to make commitments regarding resources, for instance, time, money, and people. During this process, a business need is defined, and someone in the organisation takes accountability for the project.
- ***Planning process:*** This process group is used to manage a successful project for the organisation. The planning process group establishes the project scope, and costs and schedules the activities that occur within the project. The planning process should be in line with the size of the project—that is, the larger the project, the more planning must be done.
- ***Controlling process:*** This process includes regular progress reports to ensure that the project team meets the project objectives. These objectives are in line with organisational needs to ensure that the project realises the benefits. Because of the complexity of a project, the controlling process is important to ensure that all the components are implemented as defined by the planning process. Controls inform the project team when deviations from the original plan occur, as well as measure the progress.
- ***Executing process:*** Resources are coordinated to carry out the project plan and to produce the products or services as defined in the initiating phase. The organisation must make resources available to a project to ensure its successful execution.
- ***Closing process:*** The aim during the closing process is to bring the project or life cycle phase to a systematic and orderly completion. The main focus should be to ensure that the project deliverables have been achieved and that the benefits have been realised. The formal acceptance of the project or life cycle phase is also part of this process.

6.2 Overview of Project Management Standards

There are various professional project management bodies whose sole purpose is to promote the project management discipline. This is achieved through various publications, especially project management standards. The following standards are the most well-known standards in the industry, although some countries such as Germany have their own national project management standards.

6.2.1 A Guide to the Project Management Body of Knowledge (PMI, 2017a)

A Guide to the Project Management Body of Knowledge (*PMBOK® Guide*) is a widely used standard. This standard is published by the PMI and in essence consists of 10 knowledge areas that are briefly described in the following points:

- ***Project Integration Management*** focuses on the identification, definition, assimilation, and coordination of the various processes and activities that are needed within the project life cycle as well as the process groups. This knowledge area consists of seven processes—that is, developing a project charter, developing a project management plan, directing and managing project work, monitoring and controlling project work, performing integrated change control, managing project knowledge, and the closing of either the project or a phase.
- ***Project Scope Management*** includes the processes that are required to ensure that all the work that is required to successfully deliver a project is included. The focus of this knowledge area is on what is and what is not included in the project and consists of six processes. These processes are planning scope management, collecting requirements, defining the project's scope, creating a WBS, validating the scope, and controlling the scope.
- ***Project Schedule Management*** includes the processes required to complete the project in a timely manner and consists of seven processes. These processes are planning the management schedule, defining the activities, sequencing the activities, estimating resources associated with the activities, estimating activity durations, developing a schedule, and controlling the schedule.
- ***Project Cost Management*** involves the planning, estimating, budgeting, financing, funding, managing, and controlling of the project costs to complete the project within the approved budget. This knowledge area consists of four processes—that is, cost management planning, estimating costs, determining the budget based on the costs, and controlling the actual costs.
- ***Project Quality Management*** focuses on the quality of the product or service that will be delivered by the project. The focus is on the processes and activities that determine quality policies, objectives, and responsibilities in order for the project to satisfy the needs for which is was undertaken. This knowledge area consists of three processes—planning quality management, managing quality, and controlling quality.
- ***Project Resource Management*** focuses solely on organising, managing, and leading the project team. The project team are all the people that are

assigned to the individual tasks and activities as defined in the Project Schedule Management knowledge area. Five processes constitute this knowledge area, including planning for resource management, acquiring resources, developing the team, managing the team, and controlling resources.

- ***Project Communications Management*** ensures that project information is communicated to all the team members and stakeholders in a timely and appropriate manner. It also deals with the collection, creation, distribution, storage, retrieval, and management of project information. Three processes form part of this knowledge area, starting with planning communications management, managing communications, and monitoring communications.
- ***Project Risk Management*** focuses on increasing the likelihood and impact of positive events, on the one hand, as well as decreasing the likelihood and negative impact of negative events, on the other hand. This knowledge area consists of seven processes—that is, planning risk management, identifying risks, performing qualitative and quantitative risks, planning risk responses, monitoring risks, and implementing risk responses.
- ***Project Procurement Management*** focuses on the processes that are needed for the purchasing or acquiring of products and/or services that are required from outside the normal project team. These products and/or services can be sourced either internally or externally to the organisation. Project procurement management consists of four processes—procurement management planning, conducting procurements, controlling procurements, and closing procurements.
- ***Project Stakeholder Management*** includes the processes that are required to identify people, parties, or organisations that can have either a positive or negative influence on the project itself and the project deliverables. This knowledge area consists of four processes. The first two processes focus on identifying the stakeholders, followed by planning stakeholder engagement. The remaining two processes focus on managing and monitoring stakeholder engagement.

These knowledge areas are interrelated, and the project manager must be competent in all these knowledge areas. All 10 knowledge areas are important for the success of a project, and no one knowledge area is more important than another. This interrelationship is illustrated in Figure 6-1.

The *PMBOK®* *Guide* is extremely thorough, consisting of 50 processes within the 10 knowledge areas, and each process has inputs, tools and techniques, and outputs.

Figure 6-1 *PMBOK® Guide* Knowledge Areas

6.2.2 APM Body of Knowledge (Association for Project Management, 2006)

The *APMBOK* consists of 52 knowledge areas, which are divided into the following 7 main sections:

- ***Project management in context:*** This section focuses on the placement of project management within organisations and addresses the boundaries of project management. It covers generic topics such as P3 management, the context of a project, project sponsorship, and project offices.

- ***Planning the strategy:*** This section focuses on the strategy. A plan to execute the strategy has to be developed. A well-planned project has a better chance to succeed than a poorly planned project. Seven topics are covered within this section including (i) project success and benefits management; (ii) stakeholder management; (iii) value management; (iv) project management plan; (v) project risk management; (vi) project quality management; as well as (vii) health, safety, and environment (HSE) management.
- ***Executing the strategy:*** Once the strategy has been planned, it needs to executed. The context within which the project is executed has to be monitored and controlled. Changes to the plan should be governed by formal change controls. Eight topics are covered within this section focusing on scope management, scheduling, resource management, budgeting and cost management, change control, earned value management, information management, and issue management.
- ***Techniques:*** According to the *APMBOK*, a number of techniques are used to assist in the successful delivery of a project. The project manager can decide which techniques are applicable, but those in the *APMBOK* are requirements management, development, estimating, technology management, value engineering, modeling, and testing, as well as configuration management.
- ***Business and commercial:*** The importance of the business and commercial environment cannot be negated and should be taken into account. The environment is determined either by the project itself or is imposed by the organisation's standard practices. The following aspects are covered within this section: the business case, marketing and selling of the project and its deliverables, the financing and funding of the project, procurement, and legal awareness.
- ***Organisation and governance:*** The roles and responsibilities of a project should be set out in a structured manner. This structure of roles and responsibilities covers the project life cycle as well as the organisational hierarchies. Topics include the importance of project life cycles, the identification of the need and the business case, planning the project and the creation of the project management plan, realising and delivering the plan, and putting the deliverables into operational use.
- ***People and the profession:*** The *APMBOK* stresses the importance of people and the role that they play within a project. Work is performed by people, and the success of a project is ultimately dependent upon the people within the project. Important factors highlighted by the *APMBOK* include communication, teamwork, leadership, conflict management, negotiation, human resource management, behavioural characteristics, learning and development, as well as professionalism and ethics.

Figure 6-2 *APMBOK* Main Sections

Figure 6-2 provides a graphical layout of the knowledge areas.

The last standard that will be discussed is ISO 21500 produced by ISO® itself.

6.2.3 ISO 21500:2012—Guidance on Project Management (Skogmar, 2015)

According to ISO, ISO 21500:2012 provides guidance for project management and can be used by any type of organisation and for any type of project, regardless of complexity, size, or duration. The ISO standard resembles the *PMBOK® Guide* but without the tools and techniques section. The standard consists of two main sections. The first section discusses project management concepts. There are 12 concepts, ranging from a brief discussion on projects and project management to project constraints. The second section discusses the project management processes. As with the *PMBOK® Guide*, 40 processes are divided into 10 subject groups. These 10 subject groups vaguely represent the 10 knowledge areas of the *PMBOK® Guide*. The 40 processes are divided into 5 process groups.

6.3 Methodologies

A methodology can be defined as the procedures that are followed in implementing processes in a given project (Lewis, 2000). These procedures may include forms and/or templates that must be completed, various project meetings, and change approval procedures. Bal and Teo (2001) define a methodology as a collection of procedures, techniques, tools, and documentation that should help project managers to implement a project. Phillips, Bothell, and Snead (2002) are of the opinion that a methodology is a process that is successful regardless of the scope and size of the projects, the tools used for the projects, and the people working on the projects. A methodology is a repeatable process with project-specific methods, best practices, rules, guidelines, templates, checklists, and other features for building quality systems that are manageable and deliver value to an organisation (Murch, 2005).

6.3.1 PRINCE2 (Office of Government Commerce, 2009)

PRINCE2® is a process-oriented project management methodology and consists of four integrated elements. These elements are principles, themes, processes, and the project environment itself.

- ***Principles:*** The principles are guiding obligations, and all seven principles must be performed, otherwise a project cannot be perceived as a PRINCE2 project. These principles provide a basis for good practice and can be summarised as (i) continued business justification; (ii) learn from experience; (iii) defined roles and responsibilities; (iv) manage by stages; (v) manage by exception; (vi) focus on products; and (vii) tailor to suit the project environment.
- ***Themes:*** The seven themes describe aspects of project management that should be addressed throughout the project. The seven themes are the (i) business case; (ii) the organisation itself; (iii) quality; (iv) plans; (v) risk; (vi) change; and (vii) progress.
- ***Processes:*** The purpose of the processes is to accomplish a specific objective. There are seven processes: (i) starting up a project; (ii) directing a project; (iii) initiating a project; (iv) controlling a stage; (v) managing product delivery; (vi) managing stage boundaries; and (vii) closing a project.

- ***Tailoring PRINCE2 to the project environment:*** PRINCE2 is not a one size fits all solution and needs to be tailored and customised for every new project. Tailoring will be based on the type of project and environment as well as the size of the project itself. Tailoring is not about omitting any principle, theme, or process. These elements must still all be present but adapted to cater to the needs of the project.

Figure 6-3 provides a high-level view of the four integrated themes of PRINCE2.

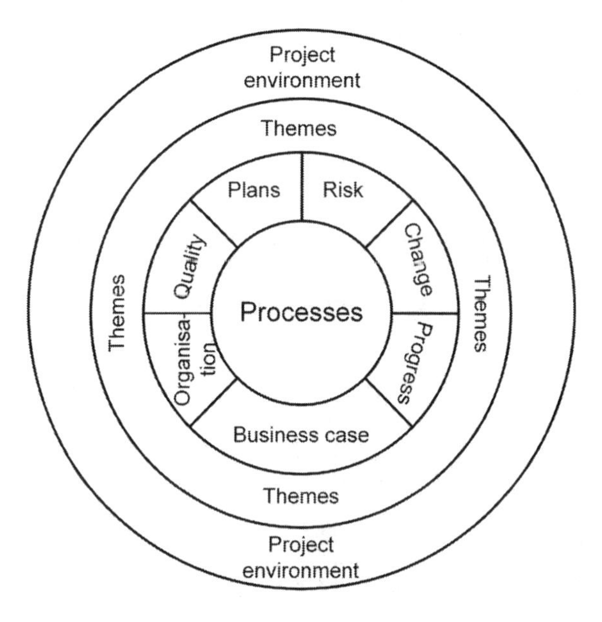

Figure 6-3 PRINCE2® Structure

Organisational leaders want and need to improve the performance of the organisation at large. One way of doing so is to improve the effectiveness of certain divisions or disciplines within the larger organisation. Project management maturity and associated models should be used to improve the effectiveness of project management within the organisation.

6.4 Talent Management

One of the components that contribute to maturity is people on the project team. The project team consists of various individuals who bring their own expertise to the team. This expertise is needed to perform certain tasks or activities that contribute to the final delivery of the project itself. One of the key players within the project team is the project manager. This section provides an overview of the competencies that a project manager must display.

Project managers should strive to achieve knowledge through education, skills through experience, and techniques through training (Mnkandla & Marnewick, 2011). It is evident that a project manager should have various competencies that are gained through various channels such as education, training, and experience. The need for a competent project manager is well documented. Kaklauskas, Amaratunga, and Lill (2010) are of the opinion that there is a relationship between achieving project success and the project managers' competencies. According to Marnewick and Labuschagne (2010), the competence of the project manager is in itself a factor in the successful delivery of projects. Although employers need guidance in the selection of a competent project manager, they are responsible for identifying the specific competencies needed for a particular project. However, the latter remains difficult to quantify.

Project management as a profession has given rise to a number of frameworks that define the scope of the discipline and describe its tools, techniques, and concepts (Chen, Partington, & Wang, 2008). Competency frameworks and standards are developed for one specific reason—that is, to assess, develop, and reassess the competencies of project managers. This process should be continuous, and project managers *per se* should embrace this process to improve their own competencies. Three major project management competency standards are discussed—GAPSS, *PMCDF*, and ICB®.

6.4.1 Global Alliance for Project Performance Standards (GAPPS)

GAPPS is an alliance of government, industry, professional associations, and training/academic institutes working together to develop globally applicable project management competency-based standards, frameworks, and mappings. The GAPPS project manager standard is written in the format of a performance-based dimension. This standard identifies six units of competency, as per Table 6-1.

The first unit of competency focuses on managing the relationships with various project stakeholders. The focus is on what is required to manage stakeholder relationships during a project (Global Alliance for Project Performance Standards [GAPPS], 2007). The aim of this unit is to demonstrate competence in ensuring that all the relevant individuals and organisations are identified in a timely and appropriate manner throughout the project lifespan.

Table 6-1 GAPPS Unit of Competency

Unit #	Unit Title
PM01	Manage Stakeholder Relationships
PM02	Manage Development of the Plan for the Project
PM03	Manage Project Progress
PM04	Manage Product Acceptance
PM05	Manage Project Transitions
PM06	Evaluate and Improve Project Performance

The second unit focuses on the competencies that are required by project managers to develop the project plan. Aspects that are included in this unit are determining the scope of the project, identification of risks, and the confirmation of the project success criteria.

In the third unit, the project manager must illustrate competence in managing the progress of the project. Elements of competence include the monitoring, evaluation, and control of the project's performance. The project manager should also exhibit competence with regard to the monitoring of risks.

The fourth unit focuses on the end product or service of the project and ensures that it is accepted by all the relevant stakeholders, as identified by PM01. The main competency is to secure acceptance of the final product or service by the stakeholders.

The fifth unit focuses on the various phases and cycles of the project. The emphasis is on the transition from one stage to another and how the project is closed at the end.

The sixth and final unit focuses on the evaluation and improvement of the project's performance. Competencies focus on the development of an evaluation plan, the evaluation of the project against this plan, and the documentation of lessons learned.

6.4.2 Project Management Competency Development Framework (PMCDF)

The *PMCDF* provides an overall view of the skills and behaviours that a project manager should be competent in (PMI, 2017b). The framework also defines the key dimensions of competence and identifies those competencies that are most likely to impact the success of a project (Pellegrinelli & Garagna, 2009). The framework consists of the following components:

- Three project management competency dimensions, defined as Knowledge, Performance, and Personal.
- The Knowledge and Performance Competence dimensions draw upon the 10 knowledge areas as well as the five project management process groups as outlined in the *PMBOK® Guide* (PMI, 2013).

The performance competencies focus on the 10 knowledge areas of the *PMBOK® Guide*. A project manager should be competent in each of the activities or processes as described in the *PMBOK® Guide*.

Personal competencies as per the PMI (2017b, p. 3) focus on the project managers with regard to their "behaviour, attitude and personality characteristics." Six units of competence are identified (PMI, 2017b):

1. Communication, which focuses on exchanging accurate, appropriate, and relevant information with stakeholders using suitable methods.
2. Leading, which guides, inspires, and motivates team members as well as the project stakeholders to achieve the project objectives.
3. The effective deployment and use of human, financial, material, intellectual, and intangible resources form part of the managing personal competence.
4. Applying the appropriate depth of perception, discernment, and judgment to effectively direct a project in a changing and evolving environment is the focus of the cognitive ability competence unit.
5. Effectiveness is the fifth competence unit, and the emphasis is on producing the desired results through the application of skills, knowledge, and tools on all the project management activities.
6. Professionalism focuses on the level of conformance to responsibility, respect, fairness, and honesty.

6.4.3 Individual Competence Baseline (IPMA ICB)

The IPMA ICB® describes the knowledge, experience, and personal attitudes expected of project managers (Caupin et al., 2006; Ghosh et al., 2012). Furthermore, the ICB deals with a mix of knowledge about project management concepts, demonstrable performance against each knowledge topic and specific behaviours that are deemed to be associated with good project management (Aitken & Crawford, 2008).

ICB organises the competence elements required by the modern project manager into 29 elements, which are organised into three competence areas (International Project Management Association [IPMA], 2015), as follows:

1. People defines the personal and interpersonal competences required to succeed in projects.
2. Practice defines the technical aspects of managing projects.
3. Perspective defines the contextual competences that must be navigated within and across the broader project environment.

The people competence area consists of 10 competencies, as follows:

1. ***Self-reflection*** focuses on the project manager's ability to acknowledge, reflect on, and understand his or her own emotions, behaviour, and values and the impact thereof on project performance. ***Self-management***, on the other hand, is the project manager's ability to set personal goals and to validate progress against these goals.
2. Project managers must demonstrate personal ***integrity*** and reliability because a lack of these qualities may lead to the failure of the intended project results.
3. Personal ***communication*** focuses on the exchange of information as well as the accurate and consistent delivery of project information to all relevant stakeholders.
4. The ***relations and engagement*** competency element focuses on the forging of personal relations that forms the foundation of productive collaboration, personal engagement, and commitment. This competency element is enforced by empathy, trust, confidence, and communication skills.
5. ***Leadership*** provides direction and guidance to individuals and groups. This competence focuses on the ability to choose and apply appropriate styles of management in different situations.
6. ***Teamwork*** focuses on building a productive team by forming, supporting, and leading the team. Team communication and team relations are among the most important competencies of successful teamwork.
7. ***Conflict*** and crisis includes the moderation and solving of conflicts through the observation of the environment.
8. The competence element of ***resourcefulness*** is the ability to apply various techniques and ways of thinking to defining, analysing, prioritising, and finding alternatives for and dealing with or solving challenges and problems.
9. Project managers must be competent in ***negotiation***. The focus is to balance different interests, needs, and expectations in order to reach a common agreement and commitment while maintaining a positive working relationship.
10. ***Results orientation*** is the critical focus maintained by the project manager on the outcomes of the project. The individual prioritises the means and resources to overcome problems, challenges, and obstacles in order to

obtain the optimum outcome for all the parties involved. The results are continuously placed at the forefront of the discussion, and the team drives toward these outcomes.

The practice competence area consists of 13 competencies, as follows:

1. ***Project design*** addresses how competent the project manager is in interpreting the demands, wishes, and influences of the organisation and translating these into a high-level project design.
2. The competence element of ***requirements and objectives*** describes the rationale of the project's existence. The focus is on the goals that need to be achieved, the benefits that need to be realised, and which stakeholder requirements are to be fulfilled.
3. ***Scope*** defines the specific focus of the project. The project manager should be competent in describing the outputs, outcomes and benefits, and the work required to produce the project's product or service.
4. ***Time*** includes the identification and structuring of all components of a project in time in order to optimise its execution.
5. The ***organisation and information*** competence element includes the identification of the various roles and responsibilities as well as the effective information exchange within the project.
6. The ***quality*** competence element focuses on how well the project is managed as well as the quality of any product that is a deliverable of the project itself.
7. Project managers must indicate competence in the ***financial*** side of the project. The financial competence element includes activities such as the estimation, planning, spending, and controlling of financial resources.
8. The ***resources*** competence element includes defining, acquiring, controlling, and developing the resources that are necessary to realise the project's outcome.
9. ***Procurement*** is the process of acquiring goods and/or services from external parties for the project itself. Project managers should be competent in purchase planning as well as contract administration.
10. The competence element of ***plan and control*** determines the project manager's ability to create a balanced plan and to execute this plan in a controlled manner.
11. ***Risk and opportunity*** includes the competencies of identifying, assessing, planning, and the implementing controls for risks and opportunities. Risk and opportunity management helps decision makers to make informed choices, prioritise actions, and distinguish among alternative courses of action. Risk and opportunity management is an ongoing process that takes place throughout the life cycle of the project.

12. The ***stakeholders*** competence element focuses on the management and engagement of all the relevant stakeholders. Project managers should, on a constant basis, revise, monitor, and act upon their interests and influence on the project.

The perspective competence area consists of five competencies, as follows:

1. The ***strategy*** competence describes how strategies are understood and transformed by the project manager into manageable elements through the use of projects.
2. The ***governance, structures, and processes*** competence element defines the understanding of and the alignment with the established structures, systems, and processes of the organisation that provide support for projects and influence the way they are organised, implemented, and managed.
3. The ***compliance, standards, and regulations*** competence element describes how the project manager complies with external and internal standards and regulations within a given country, organisation, or industry.
4. The ***power and interest*** competence element describes how the project manager recognises and understands informal personal and group interests and the resulting politics and use of power.
5. The ***culture and values*** competence element describes the project manager's approach to the influence of the organisation's culture and values on the project.

PMI's *PMCDF* and IPMA's ICB have been mainly developed along the attribute-based dimension, whereas GAPPS' standards have been mainly developed along the performance-based dimension (Bredillet, Tywoniak, & Dwivedula, 2015). The focus of assessment is the role of individuals; the knowledge, tasks, and skills required; and what they do on their jobs.

A study done in 2013 by KPMG highlighted that there is a positive correlation between project managers who follow a project management methodology and the successful delivery of a project (Barlow et al., 2013). The problem arises when one starts to analyse what constitutes project success and which factors contribute to project success.

6.5 Project Management Maturity

When one looks at the definition of the word "mature," it implies completeness in natural development or growth. When this concept of matureness or maturity is applied to project management, it implies two viewpoints: The first is that there is or was a state where project management was broken or incomplete, and the second implies that project management can heal from this incompleteness and achieve a point of total completeness. In essence, this concept implies that the project management discipline within an organisation can grow from a point of brokenness to a point where project management is perceived as complete. The question is whether there is a point of total completeness or maturity.

Project management consists of various processes, and these processes need to be performed effectively and efficiently (Backlund, Chronéer, & Sundqvist, 2014). Project management maturity thus focuses on these processes and how they can be optimised. Backlund et al. (2014) are of the opinion that organisations with higher project management maturity levels are more successful. Pasian (2014) is also of the opinion that improvement in organisational capability is a direct result of increasing the maturity of the standardised processes. Project management maturity models (PMMMs) exist to measure an organisation's maturity with regard to the project management process. This measurement is a continuous process, wherein the first step is to determine the current level of maturity, and the second step is to determine which processes can be improved. The improvements must then be implemented, and then the process starts all over again. The aim is to move from a level of incompleteness to a level of completeness.

PMMMs are one vehicle that can be used to measure the maturity of project management, and according to Jugdev and Thomas (2002), these models provide a framework within which an organisation can develop its capabilities in order to deliver projects successfully. The benefit of measuring maturity is that it sets direction and, most importantly, instils an organisational culture change that focuses on improving the way projects are executed (Backlund et al., 2014). Pasian (2014) comments that PMMMs should be used by organisational leaders to assess progress toward maturity because these models provide a systematic means to perform benchmarking and are considered to add value to organisations trying to achieve continuous improvement.

6.6 Project Success

Chapter 2 introduced the notion of project success and how it relates to the value of project management. This section discusses the current thoughts around project success and how it is defined, how project success is measured, and what organisational factors contribute to project success.

6.6.1 Project Success Perspective

Since the inception of project management, the success of a project is measured against the iron triangle of cost, time, and scope. Although this seems archaic, each and every textbook or article refers to these constraints as a basis for measuring the success of a project. This view focuses extensively on whether the project's product or service was delivered within these constraints. There are voices within the project management community calling for modern ways to view project success. The notion is that project success must be determined by certain factors, such as the long-term benefits that the product or service contribute to the organisation, whether the product or service is actually being used by customers, and, ultimately, that the product or service contributes to the strategic success of the organisation. Bannerman (2008) summarised all these new ideas and proposed that project success should be measured across five levels:

1. ***Process Success:*** This level determines whether discipline-specific technical and managerial processes, methods, tools, and techniques are employed to achieve the project objectives.
2. ***Project Management Success:*** The focus here is on the project design parameters or objectives such as time, cost, or scope.
3. ***Product Success:*** Success is measured by the main deliverable of the project itself, and the focus is on whether specifications have been met or whether the user accepted the product.
4. ***Business Success:*** Success is determined based on whether the organisational objectives that motivated the investment have been achieved.
5. ***Strategic Success:*** This level measures whether organisational expansion or other strategic advantages were gained from the project investment, either sought or emergent.

The evolution from the triple constraint as a measure for project success to a more complex measurement for project success has been slow. The way that project success is determined has changed and will continue to change in the future.

6.6.2 Project Success Criteria

The success of a project is determined by how it is measured. If a project is measured based on time and cost, then the project might achieve these criteria, but it might be failing on strategic success. It is therefore imperative that organisational leaders determine up front how the success of a project will be determined.

Continuing down the path where project success is measured across five levels, the following criteria can be used to determine project success:

1. The first grouping of success criteria focuses on the project management processes that were followed. The focus is on whether the appropriate technical and managerial processes were selected and whether these processes were integrated. This ties in with the concept of project management maturity.
2. The second grouping focuses on the triple constraint, wherein a project should be delivered within the allocated time, budget, and scope.
3. The third grouping focuses on the deliverable itself, regardless of whether it is a service or a product. The focus should be on whether all the specifications, requirements, and user expectations have been met; whether the final product or service has been accepted by the user; and if the user is actually using the product or service.
4. The fourth set of criteria focuses on business success, and the emphasis is on whether all the benefits have been realised and whether the business strategies and objectives were achieved.
5. The fifth set of criteria focuses on the organisation at large and whether the deliverable contributed to the overall development of the organisation.

It must be noted that for each criterion, realistic measurements must be in place. It might be advisable to provide upper and lower limits of measurement within which the project manager can manoeuvre the project.

6.6.3 Project Success Factors

The question that needs to be answered is what contributes to project success or what can organisational leaders do to ensure that projects are delivered successfully. Various authors have written about the factors that contribute to project success, and these are summarised as follows (Berssaneti & Carvalho, 2015; Camilleri, 2011; Joseph & Marnewick, 2014; Joslin & Müller, 2015; Marnewick, 2012; Todorović et al., 2015):

- ***Adequate handling of change:*** Projects bring change about through the product or service that is incorporated into the organisation. There are two aspects when it comes to change. The first aspect focuses on how well the organisation itself copes with the change that is introduced. The second aspect focuses on how well the project manager deals with changes within the project environment.
- ***Good communication:*** Communication is perceived as crucial to project success because all the standards and methodologies have a knowledge

area focusing on communication. Communication in a project is essential because it could resolve various issues and concerns. Communication enables the project team to understand requirements, resolve conflicts, and define the project objectives, among others.

- ***Adequate project manager competency:*** A competent project manager contributes to the success of a project. As seen in Section 6.4, the competence of a project manager is determined at various levels and dimensions.
- ***Positive executive support:*** Executives play an important role in the success of a project. They fulfil a supportive role, either as a sponsor or as a champion. As a sponsor, an executive provides political influence as well as financial support for the project. The role of the sponsor is discussed in more detail in Chapter 12.
- ***Clear business strategies:*** It was highlighted in Chapter 3 that projects are the vehicles to implement the vision and strategies of the organisation. When the strategies are ill-defined, it is almost impossible to determine the project's objectives and direction. This will surely force the project to be a failure, when it cannot deliver on the strategies.
- ***Clear requirements definition:*** The success of a project is based on whether the product or service itself meets the requirements of the users and the organisation at large. Clear requirements are achieved through proper communication and requirements engineering (International Institute of Business Analysis™ (IIBA®), 2009; Marnewick, Pretorius, & Pretorius, 2014). Clear requirements remove ambiguity and ensure that each and every team member understands what needs to done.
- ***Frequent user involvement:*** The users, among others, provide the requirements and, thus, are the origin of the requirements at the beginning of the project. This, however, does not mean that they should not be involved during the entire project life cycle. Requirements change throughout the project life cycle, which is due to organisational and environmental changes. Frequently involving the user ensures that these changes are incorporated throughout the process, not after delivery, when everyone is dissatisfied with the end product or service. The introduction of Agile principles enforces the frequent involvement of users.
- ***Appropriate formal methodologies:*** Adherence to best practices as stipulated in standards and methodologies ensures that projects are delivered more successfully than where there is no adherence to best practices. Adherence to formal standards and methodologies also allows organisational leaders to determine deviations from these best practices.
- ***Minimised scope:*** An aspect that adds complexity to a project is the scope. The larger the scope, the more complex the project and the better the chances that the project will fail. It is therefore advisable that organisational leaders keep the scope of the project to a minimum. A fine balance

exists between what constitutes the bare minimum and what are considered luxury requirements. This can only be determined through communication and clear requirements.

- ***Realistic estimates:*** When one thinks of estimates, time and cost spring to mind. Although these are some important estimates, each and every success criterion that will be used to determine the success of the project needs to have realistic estimates. It does not make sense to estimate that the project's deliverable will increase productivity by 50%, when it is obvious that a more realistic estimate would be 20%.

The relationship between project success, success criteria, and success factors is illustrated in Figure 6-4. The success of the project needs to be measured across a continuum, and each level should have each own success criteria. Factors contributing to project success need to be incorporated into the project life cycle to improve the chances of project success.

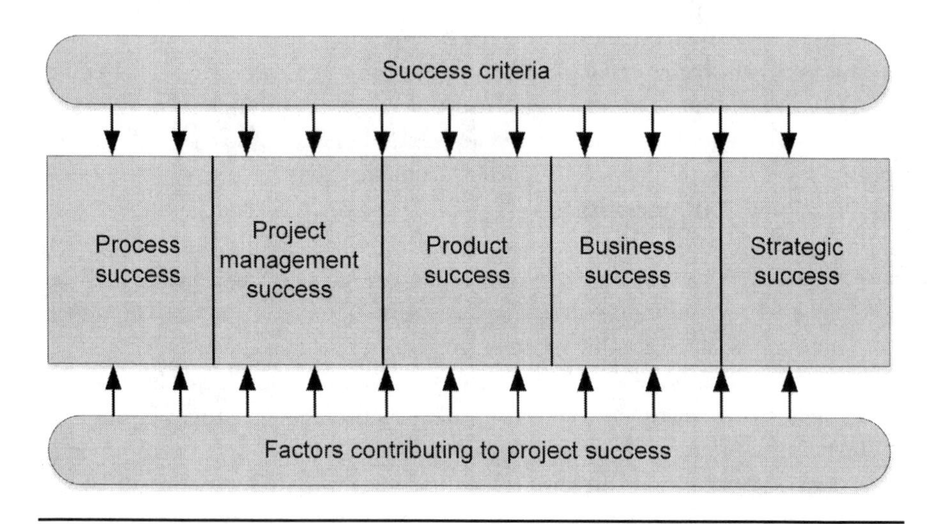

Figure 6-4 View of Project Success

6.7 The Executive's Role in Project Management

The role of the executive is prominent within project management because executive support has been identified as a factor that contributes to project success. Executives should note the importance that play in project management. The role they play in portfolio and programme management is more of a governance role, but within project management, the role shifts to a more operational role, especially when an executive is a sponsor of a project.

6.7.1 Adherence to Best Practices

Various standards and methodologies exist that can be used to manage projects within an organisation. Given the fact that project managers either have different formal project management training or are accidental project managers, each one will have his or her own ideas as to which standard or methodology to use. It is therefore imperative that the organisation adopts and adheres to one standard or methodology. This initiative should be driven from an executive level. Various factors such as the country or the type of organisation might influence this decision, but at the end of the day, project managers' performance and competencies are measured against a chosen standard or methodology. It is thus logical that an adopted standard will address various potential issues and concerns.

6.7.2 Improving Project Management Maturity (PMM)

A mature project management discipline improves the success rate of projects and ultimately the performance of the organisation. Improving project management maturity is a continuous process. Executives should encourage this quest for continuous improvement. The process described in Section 6.5 is time consuming and takes key role players away from their day-to-day activities while they are analysing the processes. Executives should support this endeavour and allow these processes to take place, or it should be part and parcel of the project managers' duties to improve project management maturity. Since PMM includes the entire organisation, executives should also provide organisation-wide support to improve it. When barriers do exist, they should be broken down by executives because project managers do not have the influence to break down these barriers themselves. Executives should have a long-term perspective in developing project management competence and skills within the organisation. Only then can PMM improve (Backlund et al., 2014).

6.7.3 Embedding Project Management as a Culture within the Organisation

Organisations are moving into an environment in which project management plays a more important role. Organisations are evolving into projectification, where more and more emphasis is on project management to accomplish day-to-day activities. This emphasis on project management is challenging because everyone needs to understand it and everything that surrounds it. This requires a totally different culture in which the emphasis is on project management and adherence to best practices. Executives should ensure that project management is embedded in the organisation and that various factors exist to accomplish this. Fernandes, Ward, and Araújo (2015) suggest 32 actions or activities that executives can follow to embed project management in the organisation and to change the culture to projectification.

6.7.4 Evolve from Triple Constraint to Strategic Success

Executives are the leaders of an organisation, and, as such, they should lead the way in which project success is measured. Executives should embrace the notion that project success is not measured anymore just on the triple constraints of time, cost, and scope. The new way of measuring success is based on a continuum of doing the project right as well as ensuring that the deliverable contributes to the overall success of the organisation. Executives should therefore enforce this new way of thinking wherein everyone who is involved in the projects determines the success of the project. Each and every project's success must be measured against predetermined criteria, and, more importantly, deviations or variations to the criteria should be allowed.

6.7.5 Project Manager Career Path

As with portfolio and programme management, a well-defined career path should exist within the organisation. This career path should be in line with the overall project management maturity of the organisation. Organisations cannot claim to be mature when most of their project managers are junior project managers. There should be a balance between junior and experienced project managers. The competencies of project managers should be well defined for each level—that is, all project managers should be competent in handling conflicts, but senior project managers should be more competent than junior project managers. Executives should therefore provide upper and lower levels of competence for each competency for each of the project management levels.

6.8 Conclusion

This chapter highlighted the importance of project management. Organisational leaders make use of projects to deliver on the vision and strategies of the organisation. It would be difficult to accomplish this without project management. This chapter presented the notions of projects and project management and the general phases and processes that are needed to manage a project successfully. These processes are generic and should be tailored to the needs of the organisation. This should be done using formal standards and methodologies. Three standards and one methodology were discussed. It seems as if the standards address the same knowledge areas, but there is a different focus or emphasis on certain aspects. It is the role of executives to determine which standard to adhere to based on what makes sense for the organisation.

Two concepts that go hand in hand are project manager competence and project management maturity. Project management competence was discussed based on three competency frameworks—GAPPS, ICB, and *PMCDF*. These three frameworks provide an overview of the competencies that project managers should master to be successful. Project management maturity (PMM), on the other hand, focuses on the organisation at large and not on the individual project manager. The focus is on where the organisation is with regard to project management and what should be in place to get the organisation to that specific point or level of maturity. The notion of project success was also discussed. The emphasis here should be on overall project success and not just on the original triple constraints. This chapter concluded with the role that executives can play in embedding project management in the organisation.

This chapter emphasises the importance of project management within the organisation. Project management *per se* is not going to achieve much if the enablers are not in place. The enablers were discussed in this chapter, and executives should focus on providing these enablers with an environment that is conducive to project management.

6.9 References

Aitken, A., & Crawford, L. (2008). *Senior Management Perceptions of Effective Project Manager Behavior: An Exploration of a Core Set of Behaviors for Superior Project Managers*. Paper presented at the Project Management Institute (PMI) Research Conference, Warsaw, Poland.

Al-Khouri, A. M. (2015). *Program Management of Technology Endeavours*. London, UK: Palgrave Macmillan.

Andersen, E. S. (2016). Do Project Managers Have Different Perspectives on Project Management? *International Journal of Project Management, 34*(1), 58–65.

Association for Project Management. (2006). *APM Body of Knowledge* (5 ed.). Buckinghamshire, UK: Association for Project Management.

Backlund, F., Chronéer, D., & Sundqvist, E. (2014). Project Management Maturity Models—A Critical Review: A Case Study within Swedish Engineering and Construction Organizations. *Procedia—Social and Behavioral Sciences, 119*(0), 837–846.

Bal, J., & Teo, P. K. (2001). Implementing Virtual Teamworking: Part 3—A Methodology for Introducing Virtual Teamworking. *Logistics Information Management, 14*(4), 276–292.

Bannerman, P. L. (2008). *Defining Project Success: A Multilevel Framework*. Paper presented at the PMI Research Conference: Defining the Future of Project Management, Warsaw, Poland.

Barlow, G., Woolley, P., Rutherford, L., & Conradie, C. (2013). *KPMG Project Management Survey Report 2013*. Retrieved from http://img.scoop.co.nz/media/pdfs/1307/KPMG_Project_Management_Survey_Report_2013_MedRes.pdf

Berssaneti, F. T., & Carvalho, M. M. (2015). Identification of Variables that Impact Project Success in Brazilian Companies. *International Journal of Project Management, 33*(3), 638–649.

Besner, C., & Hobbs, B. (2006). The Perceived Value and Potential Contribution of Project Management Practices to Project Success. *Project Management Journal, 37*(3), 12.

Bredillet, C., Tywoniak, S., & Dwivedula, R. (2015). What Is a Good Project Manager? An Aristotelian Perspective. *International Journal of Project Management, 33*(2), 254–266.

Camilleri, E. (2011). *Project Success: Critical Factors and Behaviours*. London, UK: Gower Publishing.

Caupin, G., Knoepfel, H., Koch, G., Pannenbäcker, K., Pérez-Polo, F., & Seabury, C. (2006). *ICB: IPMA Competence Baseline*, Version 3. Nijkerk, Netherlands: International Project Management Association.

Chen, P., Partington, D., & Wang, J. N. (2008). Conceptual Determinants of Construction Project Management Competence: A Chinese Perspective. *International Journal of Project Management, 26*(6), 655–664.

Fernandes, G., Ward, S., & Araújo, M. (2015). Improving and Embedding Project Management Practice in Organisations—A Qualitative Study. *International Journal of Project Management, 33*(5), 1052–1067.

Ghosh, S., Forrest, D., DiNetta, T., Wolfe, B., & Lambert, D. C. (2012). Enhance *PMBOK®* by Comparing it with P2M, ICB, PRINCE2, APM and Scrum Project Management Standards. *PM World Today, XIV*(I), 1–77.

Global Alliance for Project Performance Standards. (2007). *A Framework for Performance Based Competency Standards for Global Level 1 and 2 Project Managers*. Sydney, Australia: Global Alliance for Project Performance Standards.

Hyväri, I. (2006). Success of Projects in Different Organizational Conditions. *Project Management Journal, 37*(4), 31–41.

International Institute of Business Analysis™. (2009). *A Guide to the Business Analysis Body of Knowledge® (BABOK® Guide)* (2 ed.). Toronto, Canada: International Institute of Business Analysis.

International Project Management Association. (2015). *Individual Competence Baseline for Project, Programme & Portfolio Management*, Version 4.0 (pp. 416). Zurich, Switzerland: International Project Management Association.

Joseph, N., & Marnewick, C. (2014). *Structured Equation Modeling for Determining ICT Project Success Factors*. Paper presented at the PMI Research and Education Conference 2014, Portland, Oregon.

Joslin, R., & Müller, R. (2015). Relationships between a Project Management Methodology and Project Success in Different Project Governance Contexts. *International Journal of Project Management, 33*(6), 1377–1392.

Jugdev, K., & Müller, R. (2005). A Retrospective Look at Our Evolving Understanding of Project Success. *Project Management Journal, 36*(4), 19–31.

Jugdev, K., & Thomas, J. (2002). Project Management Maturity Models: The Silver Bullets of Competitive Advantage? *Project Management Journal, 33*(4), 4–14.

Kaklauskas, A., Amaratunga, R., & Lill, I. (2010). *The Life Cycle Process Model of Efficient Construction Manager*. Paper presented at the COBRA 2010, Paris. Retrieved from http://usir.salford.ac.uk/9697/

Kloppenborg, T. J. (2015). *Contemporary Project Management* (3 ed.). Stamford, CT, USA: Cengage Learning.

Lewis, J. P. (2000). *The Project Manager's Desk Reference* (2 ed.). Boston, MA, USA: McGraw-Hill.

Marchewka, J. T. (2009). *Information Technology Project Management* (3 ed.). Hoboken, NJ, USA: John Wiley & Sons.

Marnewick, A., Pretorius, J. H. C., & Pretorius, L. (2014). Requirements Practitioner Behaviour in Social Context: A Survey. *SAIEEE Africa Research Journal, 105*(4), 147–155.

Marnewick, C. (2012). *A Longitudinal Analysis of ICT Project Success.* Paper presented at the Proceedings of the South African Institute for Computer Scientists and Information Technologists Conference, Pretoria, South Africa.

Marnewick, C., & Labuschagne, L. (2010). A Conceptual Framework to Improve the Project Delivery Capability within an Organisation. *Acta Commercii, 10*, 249–263.

Milosevic, D. Z., & Srivannaboon, S. (2006). A Theoretical Framework for Aligning Project Management with Business Strategy. *Project Management Journal, 37*(3), 13.

Mnkandla, E., & Marnewick, C. (2011). Project Management Training: The Root Cause of Project Failures? *Journal of Contemporary Management, 8*, 76–94.

Murch, R. (2005). Methodologies in IT: Comprehension, Selection, and Implementation. Retrieved from http://www.ibmpressbooks.com/articles/article.asp?p=370635

Office of Government Commerce. (2009). *Managing Successful Projects with PRINCE2* (5 ed.). United Kingdom: The Stationary Office.

Pasian, B. (2014). Extending the Concept and Modularization of Project Management Maturity with Adaptable, Human and Customer Factors. *International Journal of Managing Projects in Business, 7*(2), 186–214.

Pellegrinelli, S., & Garagna, L. (2009). Towards a Conceptualisation of PMOs as Agents and Subjects of Change and Renewal. *International Journal of Project Management, 27*(7), 649–656.

Phillips, J. J., Bothell, T. W., & Snead, G. L. (2002). *The Project Management Scorecard.* Amsterdam, Netherlands: Butterworth-Heinemann.

Project Management Institute. (2013). *A Guide to the Project Management Body of Knowledge (PMBOK® Guide)* (5 ed.). Newtown Square, PA, USA: Project Management Institute.

Project Management Institute. (2017a). *A Guide to the Project Management Body of Knowledge (PMBOK® Guide)* (6 ed.). Newtown Square, PA, USA: Project Management Institute.

Project Management Institute. (2017b). *Project Manager Competency Development Framework* (3 ed.). Newtown, PA, USA: Project Management Institute.

Schwalbe, K. (2010). *An Introduction to Project Management.* Minneapolis, MN, USA: Kathy Schwalbe, LLC.

Skogmar, K. (2015). *PRINCE2®, the PMBOK® and ISO 21500:2012.* Retrieved from https://www.axelos.com/news/new-white-paper-explores-prince2-pmbok-iso-21500

Todorović, M. L., Petrović, D. Č., Mihić, M. M., Obradović, V. L., & Bushuyev, S. D. (2015). Project Success Analysis Framework: A Knowledge-Based Approach in Project Management. *International Journal of Project Management, 33*(4), 772–783.

Tummala, V. M. R., & Burchett, J. F. (1999). Applying a Risk Management Process (RMP) to Manage Cost Risk for an EHV Transmission Line Project. *International Journal of Project Management, 17*(4), 223–235.

Chapter 7

Project Management Offices

~ The PMO can be the battleground between empowerment and control, between people and processes, and between political factions. ~

— Pellegrinelli & Garagna*

The previous three chapters focused on the disciplines of portfolio, programme, and project management. Although there is some interrelationship between these three disciplines, it is very easy for them to function independently, especially within larger organisations.

A concept that could act as a conduit between these three disciplines is the Project Management Office (PMO). The PMO therefore plays an important role within the organisation. The first role is that of governance. The PMO should ensure that portfolio, programme, and project management are performed as intended and described in the previous chapters. The PMO also plays the role of an enabler, ensuring that all the supporting processes and structures are in place in order for these three disciplines to function optimally.

One of the concerns with regard to PMOs is that they come and go within organisations. The question raised is whether organisations really need PMOs because they add additional bureaucracy and costing. The purpose of this chapter is to highlight the importance of the PMO within the organisation. This is done first by defining the PMO and second by describing the role of the PMO.

---

* Pellegrinelli, S. & Garagna, L. Towards a Conceptualisation of PMOs as Agents and Subjects of Change and Renewal. *International Journal of Project Management*, 2009.

The maturity of the PMO is discussed in the third section because the continuous existence of the PMO is linked to its maturity level. The fourth section focuses on the performance of the PMO. The chapter concludes with the role that executives can play in the continuous functioning of the PMO.

7.1 Portfolio Management Offices

PMOs function under various names, such as the project office, the project support office, the project management centre of excellence, or even the directorate of project management. Regardless of what organisational leaders call a PMO, the functions and roles are more or less the same.

The first aspect that needs to be clarified is the definition of a PMO. The PMO, according to Aubry (2015), is the organisational entity that is assigned a variety of roles and functions in executing portfolio, programme, and project management. Hurt and Thomas (2009) are of the opinion that the PMO is a mechanism that should manage the investment in portfolio, programme, and project management. According to PM Solutions (2014), the PMO is the central organisational structure that is used to standardise the P3 practices of organisations. PMOs may also act as governance institutions, with a wide variety of roles, mandates, and even locations (Müller et al., 2013). Another view of the PMO is that it is a management structure that standardises 3P processes, and it facilitates the sharing of resources, methodologies, tools, and techniques (Jalal & Koosha, 2015). A more simplistic view is provided by Pemsel and Wiewiora (2013), who state that the PMO is just a formal layer of control between the organisational leaders and project management.

Given all these definitions, it is difficult to really determine what a PMO is, and more importantly, what the role of the PMO is within the organisation. Darling and Whitty (2016) summarise this dilemma by stating that the functions and practices expected of PMOs differ as widely as the industries and organisations that host them. This confusion contributes to a recurring theme—that is, the perceived failure of PMOs. The sustainability of the PMO is a tenuous issue (Hurt & Thomas, 2009), which is supported by data from Aubry (2015) stating that the life expectancy of a PMO is approximately two years and that only 15% of PMOs have a life expectancy of five years and more. Research done by PM Solutions indicates that the average life span of a PMO is a mere four years. The reason for the short life expectancy can be attributed to the fact that organisational leaders perceive the PMO as overhead and believe that the PMO does not really add value to the organisation at large (PM Solutions, 2014). A distinction should be made between a temporary PMO that serves the needs of a single large project or programme and an institutionalised PMO. The temporary PMO will only exist as long as the project is active. It will cease to exist with the termination of the project. The focus of this chapter is on an institutionalised PMO that has the sole purpose of enhancing the disciplines of portfolio, programme, and project management.

The physical location of the PMO is also a contentious issue as well as the number of PMOs that are allowed within an organisation. A PMO might be situated within a department or it might be situated at a more strategic level (Curlee, 2008). Regardless of the position of the PMO, it should perform the mandate that it was given. The trend is that several PMOs are established within the organisation at various levels of the organisational hierarchy (Tsaturyan & Müller, 2015). Rather than functioning in isolation, these various PMOs are interdependent. It can be concluded that PMOs are extremely heterogeneous in that PMOs vary in size, function, and the mandate that they should perform.

This confusion in the role of the PMO forces one to rethink the role that it performs within the organisation. Gartner is of the opinion that the PMO should play the role of an activist and that the focus should be more on connecting programmes and projects with the strategies of the organisation (Light, 2014).

7.1.1 Activist PMO

Activism, according to the *Oxford Dictionary of English* (*ODE*), implies the "action of using vigorous campaigning to bring about political or social change" (Stevenson, 2010). In the case of an activist PMO (aPMO), it is implied that an aPMO must vigorously campaign to bring about change. This change can be achieved through the application of the eight attributes of the aPMO. Light (2014) mentions the eight attributes of an aPMO, as per Table 7-1.

Table 7-1 Attributes of an Activist PMO (aPMO)

#	Attribute
1	Probes for risk, asks tough questions
2	Data-driven recommendations for decisions
3	Metrics for self-improvement and shared goals
4	Promotes the PMO as a source of the best project managers
5	Sets up projects with the project managers and provide on-the-job training value
6	Customises good project artefacts and shares them among the organisation
7	Works organisational politics up, down, and across the organisation
8	Embraces flexibility, enabling a dynamic structure that allows for different methodologies for different kinds of projects

Note: Data derived from Light (2014).

Unger, Gemünden, and Aubry (2012) state that PMOs should have a "strong governance mandate to guarantee the implementation of organisational goals and stake-holder interests." This governance structure should incorporate the following aspects (Unger et al., 2012, p. 611):

1. Involvement of the PMO in structuring the portfolio of project and programmes that are derived from the organisation's strategy, such as the evaluation of proposals and selection of projects. This ties in with ***attribute 1***, wherein the risk should be evaluated, and tough questions should be asked.
2. Resource management that addresses the effective and efficient allocation of limited (mainly human) resources, including cross-project (re-) allocation and formal approval. Although this seems old school and not necessarily the function of an aPMO, this notion is supported by Hurt and Thomas (2009). They propose that it is important that the PMO leader develop an understanding of the skill sets and competencies that are prevalent within the organisation and among its project managers. This is critical to building something that will be able to add value to the organisation. This ties in with ***attribute 4***, which promulgates that the aPMO "are sufficiently staffed with senior project and program managers to provide substantial—and domain-specific—mentoring and coaching" (Light, 2014, p. 6).
3. Continuous organisational (PMO) learning and portfolio exploitation that summarises tasks that cater to projects at the end of their life cycle. These tasks or activities are concerned with post-project reviews and lessons learned. This, in itself, is not conducive to learning because learning should take place during the implementation of the project/programme itself. ***Attribute 2*** plays a vital role in the continuous learning and improvement of the PMO. The purpose of the continuous learning should be to uncover the root causes of a variety of project issues as they occur and not as an afterthought.

PM Solutions found in its research (2014) that PMOs in high-performing organisations continue to be trusted with high-value, strategic responsibilities. It found that in 77% of the cases, PMOs participate in strategic planning, and 87% of the time, PMOs focus on aligning projects with strategic objectives (PM Solutions, 2014). These results contradict Gartner's claim. However, PM Solutions also coined a term—the "Disruptive PMO"—which is very much in line with the aPMO. Disruptive PMOs will need to focus more and more on strategy, innovation, agility, and stakeholder engagement. In particular, disruptive PMOs will have to develop expertise in organisational change management, which will be essential to the ultimate success of the PMO (PM Solutions, 2014).

For the PMO to provide value to the organisation at large, the following should be taken into consideration (Hurt & Thomas, 2009):

1. A core ideology for the PMO must be decided on that focuses on the long term. This ideology must realise and embrace the following:
   - Flexibility AND a standard methodology are compatible concepts
   - The PMO has the ability to be both a competent leader AND manager
   - The PMO must have both a people AND a task focus
   - The PMO must manage internal AND external relationships
   - The PMO is in the best position to support, manage and develop project managers than anyone else in the organisation

   This focus on flexibility and adaptability supports the gist of attribute 8. ***Attribute 8*** focuses on the fact that one shoe does not fit all. The PMO must adapt to its particular circumstances and do what is best for the organisation at that specific point in time.
2. The PMO must have the right leaders in place. These leaders should be passionate, focused, and determined about what constitutes effective project management. Another attribute of a right leader is a low-key, patient, but confident personality.
   - A culture of discipline must be created wherein the focus is on accountability and responsibility. This culture of discipline is inculcated by the PMO, which closely monitors, coaches, and mentors project managers in fulfilling their responsibilities. ***Attribute 5*** focuses on the role that experienced project managers play in the coaching and mentoring of inexperienced project managers.
3. PMOs should confront the brutal facts that are associated with a project and ensure that these facts are recognised and addressed in an effective and timely manner. This ties in with attributes 1 and 3. ***Attribute 3*** focuses on the collection of metrics for self-improvement. These metrics are used as a critical evaluation tool and can be used to improve the overall performance of the PMO and the project managers themselves.

Aubry, Hobbs, and Thuillier (2009) proposed that a PMO should not be considered as an isolated island in the organisation, but rather as one part of an archipelago. They also suggest that project management *per se*, and, specifically, the PMO should evolve continuously. The PMO should adapt to changes in its external or internal environment or as a response to internal tensions.

The aPMO is nothing new. It is just a re-emphasis of what a PMO should be doing and the role that it should play within a division or an organisation. The next section focuses specifically on the role of the PMO and what executives can expect of it.

7.2 The Role of the PMO

Many organisations implement PMOs without a clear direction and vision of what role they want the PMO to play (Pemsel & Wiewiora, 2013). It is therefore important that executives determine up front what role the PMO should perform in the organisation. The role of the PMO can be divided into four categories, all of which should be covered as the PMO matures into a centre of excellence. This should ensure that the PMO is sustainable and is seen as a crucial element within the organisation.

7.2.1 Things-Related Activities

The first category focuses on the details within a project and how the PMO can assist in developing a well thought through project schedule. Some of the activities that the PMO can assist with include the following:

- ***Assistance with the organisational project management standard:*** Chapter 6 discusses the various project management (PM) standards and methodologies that an organisation can adopt. The PMO can play a crucial role in assisting project managers with these standards and methodologies. It can also assist project managers with the application of the knowledge areas—for example, assisting in determining the scope and costing of a project.
- ***Assistance with monitoring and controlling:*** It was mentioned earlier that the PMO plays a governance role. The PMO can assist project managers with the monitoring and controlling of a project because the PMO will have access to all the relevant project information (Hurt & Thomas, 2009). This information can be used by the project manager to determine whether the project is deviating from the agreed-upon constraints.
- ***Change management:*** Projects bring about organisational change, and project managers do not always have the expertise or knowledge to manage the change management process. The PMO can manage this process on behalf of the project manager through communication and other related activities, such as workshops and information sessions.
- ***Performance reporting:*** An important role of the PMO is to report on the performance of the projects. Performance reporting will be based on facts because the PMO should have all the relevant data and information regarding a project. Project managers themselves will no longer report on the performance of a project, which will reduce the risk of the stakeholders receiving falsified information.

- ***Auditing:*** The PMO can play an oversight role and audit projects as they are executed and implemented. Project auditing can occur before, during, and after project completion. Governance and auditing are discussed in detail in Chapter 10. The PMO can ensure that the processes and metrics are in place to audit the project.

There might be other activities associated with this category, and it is the responsibility of the PMO and the executives to determine what activities should be performed by the PMO.

7.2.2 People-Related Activities

This second category focuses on the people aspect that the PMO should be managing. Some of the activities that the PMO can assist with are:

- ***Conflict management:*** Conflict is part and parcel of any project, and project managers should be able to resolve conflicts internal to the project. The problem arises when conflict occurs among project managers and between project managers and stakeholders. The PMO can play a mediating role in these types of conflicts because the PMO is impartial.
- ***Contract development:*** Project managers may not have legal expertise and are not in the best position to draw up legal contracts. This service can be provided by the PMO, which can either contract a lawyer as needed or employ a full-time lawyer, if the expense is warranted.
- ***Negotiations:*** The PMO should play the role of a negotiator among project managers when resources and the schedule are affected. The aim of the negotiations should be to achieve the best outcome for all interested parties, especially the organisation at large. Negotiations can also be facilitated between project managers and stakeholders.

7.2.3 Project-Focused Functions

This category of functions focuses on the project at large and not the details per se, as with the Things-Related activities.

- One of the most important functions of the PMO is to standardise the project management environment. Aspects that should be standardised is the PM standard or methodology, the PM software that will be used by the project managers, and relevant PM documentation (Hurt & Thomas, 2009).

- The PMO should also perform the functions of mentoring and consulting. Mentoring should be provided to novice project managers with regard to the application of the standards, but mentoring should also be provided to senior project managers. This places the PMO under pressure because the PMO must have the ability to mentor project managers across the board and across a variety of issues. The PMO must also act as a consultant and provide expert advice to the project managers.
- The PMO should also assist in the drafting of the business case. The PMO can provide essential information to the project manager during the creation of the business case. This assistance can be extended into the creation of a scope statement and the provision of assistance for the project start-up.
- The PMO should also be responsible for benefits tracking. This is especially important after project closure, when the project manager is assigned a new project. Processes and metrics should be in place to track the benefits and to ensure that they are realised as promised in the business case.
- The PMO should also provide basic administrative assistance to the project managers. Assistance can be provided in terms of arranging flights and accommodations, the booking of venues, and the payment of contractors.
- The PMO should also be responsible for reporting. Reporting can include the status of the projects, resource allocation, risk management, and the overall performance of the project portfolio.
- The PMO should also act as a repository for project documentation. This documentation should be available and accessible to project managers and stakeholders. It can be used for auditing purposes or as a source of information, especially in the case of lessons learned.

7.2.4 Organisation-Oriented Functions

The last category focuses on the role of the PMO with regard to the organisation and how the PMO can assist the organisation in achieving its strategies and business objectives. The functions outlined below have more long-term effects:

- ***Promoting consistency and uniformity in project management:*** The purpose of this activity is to create a long-term view of project management within the organisation. The ultimate goal is to create a project management culture in which all programmes and projects are managed according to best practices. The value of project management also needs to be continuously demonstrated to the organisational leaders in order for them to support project management and ultimately the PMO.
- ***Providing a centralised point of reference for the project management practice:*** This activity focuses on the dissemination of best practice

and state-of-the-art procedures and guidelines. It also integrates positive project best practices and promotes the use of recommended tools and techniques.

- ***Imparting specific skills and knowledge through training to project professionals:*** This activity focuses on the competence of P3 managers. The previous chapters highlighted the various competencies that P3 managers should master. A long-term strategy with regard to competent P3 managers should be determined, focusing on a gap analysis and possible certifications.
- ***Reward and recognition:*** A function of the PMO is also to reward success and to recognise competent project managers. How this is done is entirely the prerogative of the PMO. The only prerequisite is that the process of reward and recognition should be fair and transparent.

There are various groupings of roles for the PMO. For example, Hobbs and Aubry (2007) grouped the roles of the PMO into six categories: (i) Monitoring and Controlling Project Performance; (ii) Development of Project Management Competencies and Methodologies; (iii) Multi-project Management; (iv) Strategic Management; (v) Organisational Learning; and (vi) any other functions not listed in the previous five groups. Jalal and Koosha (2015) list 11 functions of the PMO, ranging from developing project management methodologies to the management of vendors and contractors.

An important role of the PMO is that it functions as a knowledge agent, which is discussed in the next section.

7.2.5 Knowledge Agent

A major risk that organisational leaders face is the loss of institutional knowledge. This occurs when employees leave the organisation and take a wealth of knowledge with them. This knowledge might be on processes and the culture of the organisation. This is also the case with project management, wherein P3 managers leave the company and take a wealth of knowledge with them.

The PMO has the potential to act as the conduit between organisational and knowledge boundaries because it connects three organisational levels: executive management, the PMO itself, and the various programme and project teams (Pemsel & Wiewiora, 2013). This places an enormous responsibility on the PMO to act as the knowledge agent. A concern is raised by Müller et al. (2013) that PMO members share knowledge with executive management and the various project teams but fail to share knowledge among themselves.

PMOs must become knowledge intensive because they play an important role in managing P3 best practices, learning from projects (both failures and successes) and improving the maturity of project management and the PMO itself (Pemsel & Wiewiora, 2013). The PMO can act as four types of knowledge agent (Desouza & Evaristo, 2006):

1. ***The supporter:*** In this role, the PMO's role is purely administrative. The PMO provides project status, identifies risks and potential issues, and maintains project archives. It reports on projects, but it does not try to influence them.
2. ***The information manager:*** The PMO stores information, which is used for evaluation purposes, as a dashboard. The PMO's function is to track and report the progress of projects and to serve as a source of information about projects and consolidated status updates.
3. ***The knowledge manager:*** The PMO is a repository of best practices, but it has no administrative responsibility. The PMO provides project expertise, mentoring, and training, and is recognised by organisational leaders as the organisation's authority on all knowledge related to P3 management.
4. ***The coach:*** The PMO's role involves both enforcement and control of knowledge sharing as well as acting as the centre for best practices and knowledge. The PMO provides a proactive and active approach to knowledge sharing and learning and focuses on strategic and organisational activities to coordinate and improve project management within the organisation.

The ultimate goal is for the PMO to act as a coach when it comes to knowledge sharing, but this role is also linked to the maturity level of the PMO. Organisational leaders cannot expect a newly established PMO to act as a coach.

7.3 PMO Maturity Stages

Hill (2007) defines five maturity stages for a PMO. These five stages are the project office, a basic PMO, a standard PMO, an advanced PMO, and, ultimately, a centre of excellence.

7.3.1 Project Office

The project office is the fundamental unit of project oversight and is created as a domain of the project manager. The project manager is responsible for the successful delivery of one or more projects. The project office provides the means to ensure professionalism and excellence in applying widely accepted principles and preferred project management practices to each project. The project office performs a project oversight role, and the overall aim is to realise the project deliverables and to deliver the projects within the project constraints. The project office performs a variety of essential project management activities, such as:

- The application of project management principles and practices
- Serving as the interface to project team performance management
- The application of policies, standards and executive decisions

7.3.2 The Basic PMO

The basic PMO is the first stage that deals with multiple project oversight and control. The basic PMO has the ability to provide aggregate oversight and control of multiple projects relative to the performance of multiple project managers. The basic PMO will have minimal staff—in some cases, just one individual assigned to build the PMO's capability. This individual might also be assigned the role of establishing a basic PMO on a part-time basis.

With a focus on controlling the project management environment, the basic PMO performs a variety of centralised project management activities, including the following:

- Having the primary responsibility for the establishment of a standard approach on how project management should be performed within the organisation
- Providing the means to compile aggregate results and analyses of project status and project progress as a basis for identifying and responding to project variations, evaluating project and project manager performance, and ensuring the achievement of project objectives
- The introduction of project management as a professional discipline, which is achieved through the prescription of applicable standards, and the designation of qualified project managers, as well as the training and empowerment of project teams

The emphasis has moved from a single project to multiple projects and how to manage these projects in the same way.

7.3.3 The Standard PMO

The standard PMO is the minimum stage for a PMO. Organisational leaders should push for this stage because the standard PMO represents the essence of a complete and comprehensive PMO. Although the standard PMO also addresses project management oversight and control, it introduces support that should optimise individual project managers' performance as well as project performance. This is the minimum stage for organisations that want to (i) implement project management as a core business competency; (ii) improve project management capability; or (iii) increase project management maturity.

The standard PMO necessitates the appointment of a full-time PMO manager and some additional full- or part-time staff members who are qualified to perform and facilitate the PMO's roles and functions. The standard PMO performs complete centralised project management oversight and control activities, with an added emphasis on introducing process and practice support in the project management environment. These activities include:

- Acting as the centre for project management
- Acting as the interface between the executive management and the project management environment
- Operating as the recognised entity that influences resource allocation on projects
- Addressing the qualifications and training of project managers

7.3.4 The Advanced PMO

The focus of the advanced PMO is on integration. The business interests and objectives need to be integrated into the larger project management environment. This integration implies a change in the organisational culture and way things are done within the organisation. Integration means the collaboration with various business units and participation in the development or adaptation of best practices and processes that are common to both the business environment and the project management environment. Apart from this role, the advanced PMO should also still perform the roles and functions of the standard PMO.

The advanced PMO is a natural evolution from the standard PMO and has increased staffing and the potential for the direct alignment of resources. The advanced PMO's staff is competent in the provision of professional and administrative resources that are required to develop, implement, and manage expanded processes, programmes, and enhanced functionality.

The advanced PMO performs comprehensive and centralised oversight, control, and support activities. This is done together with expanded functionality that represents a mature business orientation toward project management. It also provides distinct expertise in state-of-the-art project management practices and procedures.

7.3.5 Centre of Excellence

The centre of excellence is a separate business unit within the organisation and has sole responsibility for project management operations across the entire organisation. The centre of excellence assumes a strategic alignment role and guides the project management environment in its continuous-improvement efforts. Activities that should be performed by the centre of excellence include:

- Providing direction and influence for project management operations
- The establishment of a project management environment and the creation of stakeholder awareness and representation across the various business units
- The sponsoring and conducting of studies and evaluations of project management functionality and business effectiveness
- The representation of business interests in the project management environment

The progression from a basic project office to a centre of excellence does not happen overnight, and organisational leaders should allow time and commit the necessary resources to allow for the progression. Achieving the stage of a centre of excellence will also contribute to the sustainability of the PMO in the long run.

7.4 Metrics for the PMO

A crucial role for the PMO is to determine what, how, and when to measure. This section highlights the important role that PMOs play in ensuring that programmes and projects are successfully delivered, as well as the role they play in ensuring that the discipline itself is fostered and nurtured within organisations.

An important aspect about metrics is that the PMO needs something to benchmark against; otherwise, the exercise of measuring becomes futile. That is why the PMO should determine up front what standards, best practices, and success criteria will be used as a benchmark. Three major areas for measurement were identified by Hill (2007), ranging from the project management processes through to the impact of the PMO on the organisation itself.

7.4.1 Process Management and Improvement Metrics

Five process areas form part of process management and are of direct interest to the PMO.

- ***Project management methodology process:*** The PMO should develop and measure the usage of the selected project management methodology and how effective it is. Metrics that can be collected include the frequency of the prescribed activity usage and how the methodology contributes to overall project management maturity.
- ***Technical processes:*** Projects involve some level of technicality that is needed to produce the product or service. Normally, the technical processes are not part of project management and are managed outside the normal project. To ensure the alignment between project management and the technical processes, the PMO should determine a set of metrics. Metrics might include the appropriateness of the technical process and the specification of the technical tools and models.
- ***Business processes:*** On the other side of the coin, project management and business processes need to be aligned. Project management forms part of the larger organisation and should be aligned to the business processes. To determine alignment with the business processes, the PMO can use some of the following metrics: scope and contract acceptance criteria, change management criteria, and vendor selection criteria.
- ***Resource management processes:*** The PMO is responsible for resource management, especially in the case of the last three stages indicated above. The PMO should communicate and collaborate with the human resources division when it defines human resource–related processes. Once again, the human resource processes of the project environment cannot be out of sync with that of the organisational human resource processes. Some metrics include resource-assignment responsibility and resource-sourcing criteria.
- ***PMO support processes:*** The PMO also needs to determine how effectively it manages its own processes. Some metrics that can be used to measure the effectiveness of the PMO include the PMO's staff workload levels as well as the frequency and content of these reports.

The second area of measurement is the performance of programmes and projects themselves. The PMO needs to determine metrics for project performance.

7.4.2 Project Performance Metrics

The focus of the project performance metrics is based on 5 of the 10 knowledge areas delineated in the *PMBOK® Guide* (PMI, 2017) but can be extended to include all of the knowledge areas.

- ***Budget:*** The PMO should provide metrics for the development and management of the project budget. The metrics should cover the entire project management process, ranging from initiation to closing. An example of a metric is how actual costing differs from the planned costing. Budget-related metrics ensure that the project remains within the allocated budget or that corrective action can be taken when metrics indicate that there is a problem.

- ***Schedule:*** As with the budget, the PMO should provide metrics ensuring that the project is managed within the project schedule. Metrics play a two-way role in the schedule. First, they can be used as guidance—for example, the preferred depth of activity levels of the work breakdown structure (WBS). Second, the metrics can be used as a controlling measure to determine whether a project is behind schedule, in which case corrective action can be taken.
- ***Resource management:*** Metrics for resource management should focus on the entire human resource knowledge area. The purpose of the metrics should be to optimise the performance of the resources. Metrics include resource utilisation, resource performance, resource acquisition, as well as team cohesion. All of these metrics should be consistent and in line with the organisation's human resource mandate.
- ***Risk management:*** Risk management is inherent to project management, and the PMO should have metrics in place to manage project risks. Guidance metrics include the frequency of risk assessment and analysis, risk impact on project performance, as well as the cost of risk mitigating. Measurement metrics should assist the PMO in its oversight role. Metrics included in this role are risks that actually happened or the number of risks associated with specific projects or individual team members.
- ***Quality assurance:*** The PMO plays an important role in the quality of a project. Quality assurance ties in with the technical processes of the project. Executing the technical processes according to specification should ensure a quality product or service. The PMO is responsible for putting metrics in place to determine the overall quality of the processes as well as the deliverable. Metrics could include adherence to quality standards criteria and the measurement of deviations, as well as the number of change requests as a consequence of defects.

The third area focuses on the business, with the purpose of integrating the business metrics into the project environment and providing accurate information to the project stakeholders.

7.4.3 Business Management Metrics

Four key project management elements form part of business management metrics.

- ***Contract and agreement metrics:*** Metrics should be provided by the PMO ensuring that contracts achieve the organisational strategies and business objectives. Metrics include pricing strategies and vendor prequalification.
- ***Customer satisfaction metrics:*** Projects are executed for customers within, or sometimes outside of, the organisation. The PMO should determine the level of customer satisfaction with the end product or service. Apart from this, the PMO should also determine the level of satisfaction with the entire project management process as well as interactions with the PMO.
- ***Project portfolio management metrics:*** Programmes and projects form part of the portfolio. The portfolio's purpose is to deliver the vision and strategies of the organisation. The PMO is in the ideal position to provide metrics and measurements of the achievement of the vision and strategies. Metrics can also be used to determine which programmes or projects should be incorporated into the portfolio and how each individual programme and project is performing.
- ***Financial metrics:*** The focus of this element is not on individual programmes or projects but on the larger financial aspect of the PMO and the P3 discipline. The focus is on the value of the PMO itself and the value of project management to the organisation.

The metrics provided by Hill (2007) can be used as guidance by executives and the PMO. A more pragmatic approach might be to take the roles and functions discussed in Section 7.2 and assign metrics to each of them. The next section focuses on the challenges that PMOs face.

7.5 PMO Challenges

As indicated earlier, PMOs do not last long in organisations, and they come and go. Only a few PMOs manage to deliver value to the organisation at large. Table 7-2 highlights the challenges that PMOs face within an organisation. Three of the top five challenges appear every time the survey is done, but two new challenges emerged for 2016.

In the first challenge, the PMO is perceived as a policeman who adds additional processes to project management, which is already burdened with processes. Adding additional PMO processes, for whatever reason, is perceived as overkill, and, therefore, people resist these processes.

The second challenge is that one of the roles of the PMO is to bring about change. This is also the case for the aPMO. By nature, people resist change, and they will resist the changes that occur because of the PMO. Resistance might

Table 7-2 Trend of PMO Challenges

2014[1]	2016[2]	Position
Organisational resistance to change	PMO processes seen as overhead	↑
PMO processes seen as overhead	Organisational resistance to change	↓
Having enough time/resources to devote to strategic activities	Demonstrating the added value of the PMO	↑
Demonstrating the added value of the PMO	Assuring the consistent application of defined processes	NEW
Inadequate resource management capability	Project managers with inadequate project management skills	NEW

[1] Data derived from PM Solutions (2014).
[2] Data derived from PM Solutions (2016).

be against the PMO itself because it institutes different ways of doing project management. Regardless, organisational leaders should accept that PMOs bring change about and that this change is for the better of the organisation.

The third challenge, which recurs consistently, is that of showing the value of the PMO. Chapter 2 dealt with the notion of project management value. Just as P3 managers must indicate the value of portfolio, programme, and project management, the PMO must make a concerted effort to highlight the value that it adds to the organisation. The PMO must advocate the victories that it achieved as well as how the project management discipline has matured under its auspices.

The fourth challenge ties in with the first challenge, wherein processes are not just perceived as overhead but also are not applied consistently. This boils down to a governance issue, and the PMO will quickly lose face if certain individuals, project managers, or executives are treated differently by the PMO.

A fifth challenge concerns the competency levels of project managers. Project manager competencies were discussed in detailed in Section 6.5. The PMO can have the best ideas and strategies to improve project success rates and mature project management. However, this will all be in vain if the executers of these ideas—the project managers—lack the required competencies and do not display the right level of skills to realise them.

In conclusion, various challenges exist that might hamper the sustained performance of the PMO. What is important is that the PMO must identify potential challenges and address these challenges in a swift and effective manner.

7.6 The Executive's Role in the Project Management Office

Executive management can play a vital role in establishing a PMO that can eventually function as a centre for excellence that adds value to the organisation and to the project management discipline itself. The following strategies can be implemented by executives to ensure a sustainable PMO.

7.6.1 The Need for a Centre of Excellence

The need for a centre of excellence is to meet future demands of the PMO as well as project management. This centre of excellence will play an important role in the achievement of the vision and strategies of the organisation. As seen in Section 7.3.5, the focus is to mature both the project management discipline as well as the PMO itself. Without the long-term vision to establish a centre of excellence, the PMO is doomed to failure if it is not part of the vision (Al-Khouri, 2015). It will then be perceived as a burden and something that does not add value to the organisation.

7.6.2 Establishing of a Vision and Values

According to Hurt and Thomas (2009), the long-term survival of the PMO is dependent on the vision for the PMO and the values that the PMO will display. The PMO must be passionate about project management and anything related to project management. The PMO should instil the perception that it is the only entity within the organisation that has this passion for project management. The PMO should survive leadership changes within the organisation and the PMO itself and adapt to changes in the needs of the organisation. Executives can support this passion of the PMO by allowing only the PMO to be the master of project management. No other entity or individual can perform this role of a passionate custodian. However, this does not mean that advice or knowledge from outside the PMO cannot be incorporated.

7.6.3 PMO Leadership

Various roles exist within the PMO, such as the PMO executive or the project management mentor (Kendall & Rollins, 2003). The leader of the PMO will play a crucial role in the sustainability of the PMO. Hurt and Thomas (2009) indicate that the PMO leader should have three qualities—that is, passion, confidence, and the ability to demonstrate the value of the PMO through quick wins. The PMO leader must be able to convince organisational leaders of the value of the PMO as well as change the way project managers work. Only a highly successful leader can achieve this responsibility. It is the role of the executives to find and appoint such a person in charge of the PMO. Apart from finding and appointing this individual, executives must also support the individual in all endeavours that are undertaken to establish a centre of excellence.

7.6.4 Staffing of the PMO

The people employed by the PMO who must enhance the project management discipline are important for the long-term existence of the PMO. These people must be knowledgeable about project management but should not take over the role of the project manager. This is a very fine line that should not be crossed. The moment that project managers believe that the PMO members want to manage and interfere in their projects, the PMO will lose its credibility. This implies that project managers will turn their backs on the PMO and will not make use of its services. Although executives do not have a direct say in who should be appointed by the PMO, they can play an oversight role, making sure that there is a clear distinction between the roles of the PMO staff and the roles of the project managers. The knowledge and skills of the PMO staff are also important in the success of the PMO. If a staff member fulfils the role of a coach, for example, then the staff member must be skilled as a coach. Each speciality role within the PMO must be filled by experts.

7.6.5 Creating a Culture of Discipline

Discipline is essential for the success of the PMO. Discipline can be divided into two parts. The first part goes hand in hand hand with the previous section, in which there is a clear delineation of duties and responsibilities. No one should tread on someone else's domain. This calls for self-discipline, especially from PMO staff when they see that a project manager is heading for disaster. The impulse might be to step in and save the project, but the discipline should be there to allow the project manager to make mistakes and to learn from them. Second, the organisation at large should show discipline and allow only the PMO to perform project management–related activities. The PMO is the first call when it comes to project management, and departments should refrain from creating their own project management environments that are isolated or contradict the values of the PMO. Executives play an important role in ensuring that the PMO is perceived as the only legitimate entity that can perform project management duties.

7.6.6 Determining and Communicating the Value of the PMO

One of the recurring challenges that PMOs face is that organisational leaders do not see the value of the PMO. Executives play an important role in advocating the value of the PMO. First, executives should determine up front how the value of the PMO will be determined. These criteria and associated metrics need to be established in partnership with the PMO leadership. If the PMO leadership understands what is expected of them, then it is easier to deliver on these expectations. Second, executives should hold the PMO accountable for achieving these criteria. The PMO, as part of its roles and functions, should continuously determine whether it still adds value to the organisation at large. Executives should also perform a control function and hold the PMO accountable if it is not achieving these criteria and could not deliver on the value aspect. But, when value is added to the organisation, executives should acknowledge this fact and communicate it to the organisation at large.

7.7 Conclusion

This chapter focuses on the project management office (PMO) as a centre of excellence that can be used by organisational leaders in achieving project management maturity and excellence. The concept of the PMO is discussed to create a general consensus about the PMO. The activist PMO (aPMO) is also included, indicating that the role of the PMO not only stays the same over the years but also needs to be altered to allow for new developments in its respective organisation. The role of the PMO is discussed, and from this discussion, it is evident that the role of the PMO is dependent on two notions: the maturity level of the PMO and the expectations of the organisation at large. The roles vary from things- and people-related activities through project-focused and, ultimately, organisation-oriented functions.

The PMO maturity stages offer insight into how organisational leaders can go about establishing a sustainable PMO that adds value to the organisation. It was highlighted that this is a long-term process in achieving the centre of excellence stage. The metrics associated with the PMO are discussed to highlight the important role that the PMO plays in creating a project management–oriented organisation. PMOs do face challenges, and a section focuses on recurring challenges that PMOs face and how executives can assist the PMO in overcoming these challenges. The chapter concludes with the role that executives can play in ensuring that the PMO is sustainable and contributes value to the organisation.

Organisational leaders can continue to manage the disciplines of portfolio, programme, and project management as separate but related disciplines. There is nothing wrong with managing in this way, but when organisational leaders wants to harness the full effect of portfolio, programme, and project management, then the implementation of a PMO is perceived as the answer. The PMO allows for the maturity of the discipline at large but also plays a crucial role in achieving the organisational vision and strategies as well as managing the benefits associated with the various programmes and projects.

The true value of portfolio, programme, and project management can only be realised when the PMO is a pivotal entity within the organisation and has the support of the executives. This allows the PMO to fulfil its mandate and really be a centre of excellence.

7.8 References

Al-Khouri, A. M. (2015). *Program Management of Technology Endeavours*. London, UK: Palgrave Macmillan.

Aubry, M. (2015). Project Management Office Transformations: Direct and Moderating Effects That Enhance Performance and Maturity. *Project Management Journal, 46*(5), 19–45.

Aubry, M., Hobbs, B., & Thuillier, D. (2009). The Contribution of the Project Management Office to Organisational Performance. *International Journal of Managing Projects in Business, 2*(1), 141–148.

Curlee, W. (2008). Modern Virtual Project Management: The Effects of a Centralized and Decentralized Project Management Office. *Project Management Journal, 39*(S1), S83–S96.

Darling, E. J., & Whitty, S. J. (2016). The Project Management Office: It's Just Not What It Used to Be. *International Journal of Managing Projects in Business, 9*(2), 282–308.

Desouza, K. C., & Evaristo, J. R. (2006). Project Management Offices: A Case of Knowledge-Based Archetypes. *International Journal of Information Management, 26*(5), 414–423.

Hill, G. M. (2007). *The Complete Project Management Office Handbook* (2 ed.). Boca Raton, FL, USA: CRC Press.

Hobbs, B., & Aubry, M. (2007). A Multi-Phase Research Program Investigating Project Management Offices (PMOs): The Results of Phase 1. *Project Management Journal, 38*(1), 74–87.

Hurt, M., & Thomas, J. L. (2009). Building Value through Sustainable Project Management Offices. *Project Management Journal, 40*(1), 55–72.

Jalal, M. P., & Koosha, S. M. (2015). Identifying Organizational Variables Affecting Project Management Office Characteristics and Analyzing Their Correlations in the Iranian Project-Oriented Organizations of the Construction Industry. *International Journal of Project Management, 33*(2), 458–466.

Kendall, G., & Rollins, S. (2003). *Advanced Project Portfolio Management and the PMO*. Boca Raton, FL, USA: J. Ross Publishing.

Light, M. (2014). *How an Activist PMO Helps Share Strategic Vision to Optimize the Portfolio.* Retrieved from https://www.gartner.com/doc/2924418/activist-pmo-helps-share-strategic

Müller, R., Glückler, J., Aubry, M., & Shao, J. (2013). Project Management Knowledge Flows in Networks of Project Managers and Project Management Offices: A Case Study in the Pharmaceutical Industry. *Project Management Journal, 44*(2), 4–19.

Pemsel, S., & Wiewiora, A. (2013). Project Management Office a Knowledge Broker in Project-Based Organisations. *International Journal of Project Management, 31*(1), 31–42.

PM Solutions. (2014). *The State of the Project Management Office (PMO) 2014.* Retrieved from http://www.pmsolutions.com/reports/State_of_the_PMO_2014_Research_Report_FINAL.pdf

PM Solutions. (2016). *The State of the Project Management Office (PMO) 2016.* Retrieved from http://www.pmsolutions.com/resources/view/the-state-of-the-project-management-office-pmo-2016/

Project Management Institute. (2017). *A Guide to the Project Management Body of Knowledge (PMBOK® Guide)* (6 ed.). Newtown Square, PA, USA: Project Management Institute.

Stevenson, A. (2010). *Oxford Dictionary of English.* Oxford, UK: Oxford University Press.

Tsaturyan, T., & Müller, R. (2015). Integration and Governance of Multiple Project Management Offices (PMOs) at Large Organizations. *International Journal of Project Management, 33*(5), 1098–1110.

Unger, B. N., Gemünden, H. G., & Aubry, M. (2012). The Three Roles of a Project Portfolio Management Office: Their Impact on Portfolio Management Execution and Success. *International Journal of Project Management, 30*(5), 608–620.

Chapter 8

Benefits Realisation Management

~ It does really need a shift of focus by organizations. Both the executives and project managers need to think about benefits as distinct from deliverables. ~

— Gary McCalden, PMO Director*

The notion of value was discussed in Chapter 2. The question was asked whether projects and project management add value to the organisation. The answer to this question actually lies in the benefits that organisational leaders harvest from the various programmes and projects that are implemented within the larger portfolio.

Benefits realisation management (BRM) is the final link between the organisational strategies and the projects that are implemented. The business case that is used to motivate a project promise that some benefits will be realised when the project is completed. At the same time, projects are used to implement the vision and strategies of the organisation. The implication is that the vision and strategies are actually achieved through the realisation of the promised benefits (Marnewick, 2014). The important issue is that programmes and projects create deliverables. Through the actual implementation and usage of these deliverables, benefits are created for the organisation.

The main focus of this chapter is to highlight BRM and the important role that it plays within the organisation. BRM should form an integral part of the

* Cooke-Davies, T. *The Strategic Impact of Projects—Identify Benefits to Drive Business Results*, 2016. Retrieved from http://www.pmi.org/learning/thought-leadership/pulse/identify-benefits-business-results

project management culture of the organisation rather than be treated as an additional process at the end of a programme or project. This chapter focuses first on BRM and the important role it plays. Second, the BRM process is discussed, including how it relates to the programme and project management processes. The notion of a benefits dependency network (BDN) and associated benefit profiles are discussed in the third part of the chapter. The chapter then concludes with a section on the role of executives in BRM.

8.1 Defining Benefits Realisation Management

Of importance is distinguishing between the concepts of benefits, benefits realisation, and benefits realisation management. The deliverable of any project brings change. When this change is perceived as positive by any of the stakeholders, then it is classified as a benefit (Bradley, 2010). The larger the change, generally, the larger the benefit. The opposite is also true where a change can be perceived as a disbenefit, implying that the change is detrimental to the organisation. The Project Management Institute (PMI) (2013) defines a benefit as an outcome that adds value or change. It seems as if there is a general consensus that a project's deliverable bring change about that should be beneficial and add value to the receiving organisation. A project can have multiple benefits associated with it—the more benefits that are associated with a project, the better. Benefits realisation *per se* focuses on the realisation of the benefits. The PMI (2013) states that some benefits are relatively certain and easy to quantify, whereas other benefits are more difficult. The purpose of benefits realisation is to maximise the likelihood that a benefit will be realised, and to ensure that all of the endeavours that are accompanying the effort to realise the benefit are in place to make it happen. BRM includes all the activities that are required to manage benefits from the moment they are identified up to the point at which they are realised and sustained (Aguilera, 2016).

BRM is becoming an important concept within the P3 discipline. Traditionally, benefits were perceived as part of programme management. This is echoed in the definitions of programme management (PMI, 2013; Sowden, 2011). Projects within a programme will deliver at least one benefit, but these definitions imply that it is almost impossible for a single project to deliver benefits on its own when it is not part of a programme. The success of a project is determined across five levels, as discussed in Chapter 6. The ultimate goal of a project is to deliver value, which can only be achieved if it is beneficial to perform the project. It does not make logical sense to implement a project if there are no benefits attached to the project. Practitioners and academics have come to the conclusion that BRM plays an important role in both programme and project management (Marnewick, 2016).

The BRM process that is discussed in the next section is therefore applicable to projects within programmes as well as single projects.

8.2 The Benefits Realisation Management Process

The PMI (2016) simplified the BRM process into three main activities, as follows:

- ***Identify benefits:*** This activity focuses on identifying potential benefits that will add value to the organisation.
- ***Execute benefits:*** The focus should be to achieve the maximum benefit through the minimisation of risks and the maximization of opportunities.
- ***Sustain benefits:*** The focus here is to ensure that the project deliverable continues to deliver value to the organisation. Organisations should therefore investigate into the enablers that will ensure the sustainment of benefits.

This process is a little bit oversimplified, but it provides organisational leaders with a quick overview of what needs to be done to identify, manage, and sustain benefits. A generic BRM process is illustrated in Figure 8-1. This can be used by organisational leaders to manage the BRM process and to ensure that the maximum value is achieved through the various deliverables.

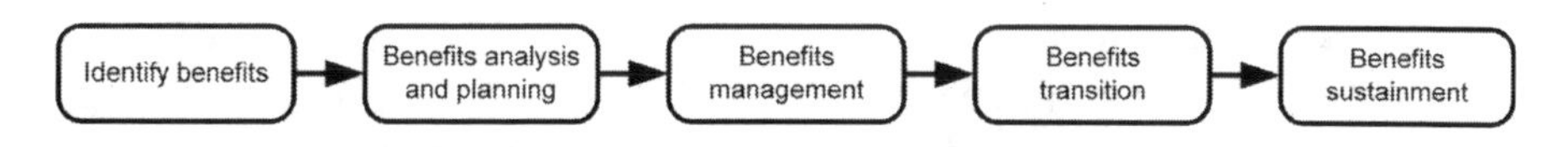

Figure 8-1 Benefits Realisation Management Process

1. ***Identify benefits:*** The first step is to identify the benefits that are associated with a project. In a programme, different projects might have the same benefit. In this case, these projects should be merged into one benefit to (i) optimise the impact of the benefit; and (ii) simplify the remainder of the BRM process. A project's business case generally defines the promised benefits and can be used to identify the benefits during this step. Part of this identification step is to identify the changes that are associated with the benefit (Bradley, 2010). When benefits are transitioned into the organisation, change is inevitable. Change might be as a result of the new deliverable, but it can also be due to changes in processes or the way people work.
2. ***Benefits analysis and planning:*** Once an inventory is created of all the benefits, the next step is then to analyse each of these benefits and plan for their realisation. The outcome of this step is a benefits realisation plan

(BRP). During the analysis phase, each identified benefit is scrutinised. The focus is on specific aspects, such as how difficult will it be to realise the benefit, how difficult will it be to transition and sustain the benefit, and what is the impact of the benefit to the organisation at large. The purpose of this step is determine a portfolio of benefits. The same logic can be applied within portfolio management, where benefits need to be prioritised. This is easier said than done. As benefits are an integral part of programmes and projects, the entire project portfolio needs to be balanced based on the promised benefits. This is the second part of this step. Once all the benefits and associated projects are analysed, the realisation of the benefits needs to be planned. The BRP determines when benefits will be realised and who is ultimately responsible for the realisation. The BRP focuses not only on the benefits but also on other activities that are required to assist in benefits realisation (PMI, 2013). The BRP should be incorporated into the overall project plan.

3. ***Benefits management:*** As with any plan, the BRP needs to be executed (Ward & Daniel, 2008). This forms part of the benefits management step. The overall purpose of this step is to monitor the progress of each benefit's realisation. First, the person responsible for the BRP should determine the progress of each benefit's realisation. This is determined at certain predefined intervals. During normal project execution, the environment and certain constraints may change. This will have an impact on benefits realisation and might result in (i) the fact that certain benefits cannot be realised anymore; or (ii) that the impact of the benefit needs to be adjusted. The benefit owner should be able to take corrective action during this step to ensure the optimum realisation of benefits.
4. ***Benefits transition:*** Benefits are realised and transitioned throughout the project life cycle, not just during the closing phases of a project. Benefits are associated with the deliverables of a project. Once the deliverables are being used by the organisation, the associated benefits are realised. The aim of this step is to ensure that the benefits are transitioned into the various operational divisions or units of the organisation. Once the benefits are utilised, their value is created for the organisation. Part of benefits transition focuses on the organisational changes that are necessary to receive the benefits and to ensure that the benefits are incorporated into the day-to-day running of the organisation. This is achieved when the project team has the opportunity to share information about how the project's deliverables can reach full operational status (Phillips, 2016). Organisational leaders eventually become the key stakeholders because they need to prepare themselves and the organisation for this inevitable change.
5. ***Benefits sustainment:*** The last step is to sustain the benefits as long as possible to maximise the value that is derived from the benefits. The

emphasis of this step should be on the activities or processes that are needed to sustain the benefits. The focus of this step is on the following (PMI, 2016): (i) the optimisation of benefits, regardless of whether the benefits are tangible, intangible, short term, or long term; (ii) the transitioning of the benefits to the appropriate stakeholders, which also includes appropriate accountability; (iii) the handing over of the project's deliverables and capabilities to the various business owners; (iv) the continuous measuring of the realised benefits, including the timeframe within which the benefits are realised; and (v) whether unanticipated benefits have been realised and captured.

The BRM process is neither a stand-alone process nor is it just an add-on process. This process needs to be incorporated within the overall project management processes and should be managed according to best practices (PMI, 2013; Sowden, 2011). The following question then arises: Who is responsible and accountable for BRM? The next section focuses briefly on the roles and responsibilities within BRM.

8.3 Roles and Responsibilities within BRM

According to Aguilera (2016), one of the most important reasons that benefits realisation management fails is the fact that no one actually knows who is responsible for the oversight role. There are two parts that need oversight. The first is, Who is responsible for constructing the benefits realisation plan? Second, and this is more difficult to answer, Who is responsible for sustaining the benefits? This becomes more difficult, especially when benefits are realised during the project and not just after the completion of the project.

Five distinct roles are identified that play a role in the successful realisation of benefits—that is, the sponsor; the benefits or business owner; and then the portfolio, programme, and project managers.

- ***Executive sponsor:*** The executive sponsor's only responsibility is to ensure that the project produces maximum value to the organisation. This is achieved when the project's deliverable realises benefits that, in turn, create value to the organisation. It implies that the sponsor should take ultimate accountability for BRM and ensure that all the enablers are in place to maximise benefits realisation and sustainment.
- ***Benefits or business owner:*** A project's deliverable is implemented by a specific business unit for a specific business owner who saw some initial value in having such a deliverable. Using the deliverable will create benefits for that business unit. It is the business owner of that business unit

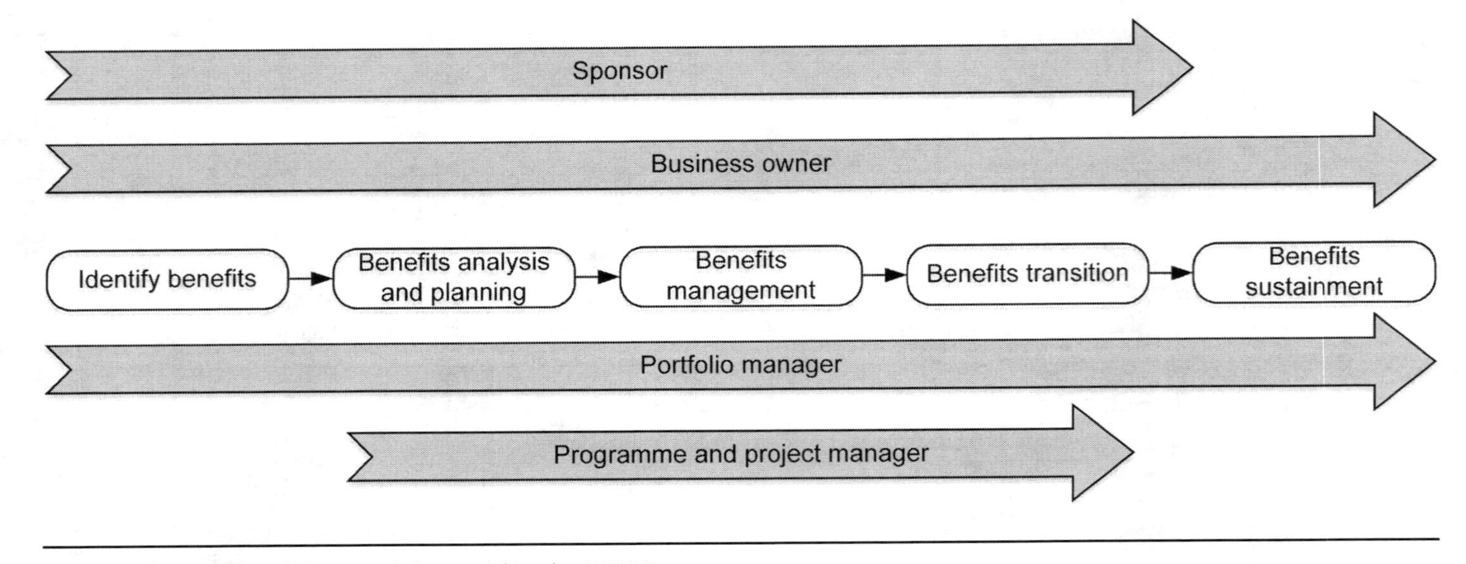

Figure 8-2 Roles and Responsibilities within the BRM Process

or division who should take overall responsibility for the monitoring and measuring of the benefits, thus ensuring that they are realised and sustained. The business owner should ensure that the deliverable is utilised to the maximum to harvest the full extent of benefits associated with the deliverable.

- ***Portfolio manager:*** The portfolio manager is responsible for establishing, balancing, monitoring, and controlling the components that form part of the portfolio. One of the aspects that should be taken into consideration when a portfolio is assembled is the benefits that are associated with any given component. Figure 4-3 highlights the portfolio management process, and during the evaluation and selection activities, benefits should also be taken into consideration. The role of the portfolio manager is to ensure that (i) benefits realisation is part of the portfolio management process; and (ii) benefits are actually realised to improve the success of the portfolio through value creation.
- ***Programme and project managers:*** These two roles are responsible for the actual drafting of the benefits realisation plan as well as the implementation of these plans. They play an active role during the second, third, and fourth steps of the BRM process. They must also ensure that the organisation is in a position to take ownership of the project deliverable and the associated benefits.

BRM is the responsibility of everyone who is involved with programmes and projects. The accountability for realising and sustaining the benefits actually lies with the business owner. Organisational leaders should ensure that everyone understands the BRM process and their associated responsibilities. Figure 8-2 maps the various roles and responsibilities within the BRM process.

Figure 8-2 highlights that although responsibilities overlap during the BRM process, every role has a distinct responsibility within this process. The different role players need to understand what a benefit is in order for them to manage it. It is easy to say that they should be able to identify, manage and sustain benefits but often it is unclear what constitutes a benefit and its dimensions. Even when the roles and responsibilities are clearly defined, challenges are still faced that hamper the management of benefits.

8.4 Challenges Facing Benefits Realisation Management

Realising benefits is not as easy as it seems because there are many hurdles that need to be crossed before the BRM process is imbedded into the organisation. It is evident that when the project management and BRM processes form a symbiotic relationship, the chances increase for project investment success (Badewi, 2016). Some of the concerns that are raised that are perceived as challenges are briefly outlined below:

- Some organisations do not have a BRM process in place, and when they do, it is very informal and not institutionalised (Bennington & Baccarini, 2004; Breese, 2012).
- The business case is primarily used to gain funding from the organisation (Marnewick, 2014). This results in the fact that benefits are often overstated in the business case for the sole purpose of gaining approval (Bennington & Baccarini, 2004; Chih & Zwikael, 2015).
- The baseline and target values of a benefit must be predicted. A period of three to four years might elapse before the actual delivery of the benefits. This makes it extremely difficult to predict benefits realisation (Bennington & Baccarini, 2004). This then results in poorly formulated project target benefits (Chih & Zwikael, 2015).
- Another concern raised by Dhillon (2005) is that organisational leaders do not know or understand how to measure the value of benefits once they are delivered, and it is difficult to assess benefits once the project is completed.
- Another challenge is to quantify intangible benefits. This is especially the case when decisions must be made based on these intangible benefits (Aguilera, 2016). Examples of intangible benefits are the brand image of the organisation or market share. Tools such as comparative or scenario analysis can be used to quantify intangible benefits. In the case of a brand image, a proprietary tool called "The BrandAsset Valuator" can be used to measure the four dimensions of a brand's image.
- BRM adds overhead to an existing project because it requires resources to manage the benefits. Organisational leaders are sometimes reluctant to allow for this overhead and do not see the benefit of having a BRM process imbedded into the organisation (Paivarinta, Dertz, & Flak, 2007).

The next section discusses the three major dimensions of a benefit.

8.5 Profile of a Benefit

A benefit comprises three major dimensions: (i) the details of the benefit itself; (ii) the various organisational processes that a benefit requires; and (iii) the necessary information that is required to track the benefit throughout the BRM process.

- ***Benefit details:*** This dimension provides general information about a benefit and includes, at a minimum, the following:
  - *Number:* Each benefit should have a unique identifier that can be used to identify it. The rationale is that more than one of the same type of benefit description can occur within a portfolio, and a unique number makes it easier for identification purposes.
  - *Description:* Each benefit should have a short description about what the benefit entails. This must be concise but yet descriptive.
  - *Organisational objectives supported:* Projects are aligned with organisational strategies and objectives as is the case with benefits. During the identification step, each benefit must be mapped to the organisational strategies and objectives and to what extend these objectives are supported by the benefit. It might happen that a benefit supports one or more objectives.
  - *Other benefits to which this benefit contributes:* Some benefits are not realised as single concerns but might be related to other benefits. This is especially the case within a programme. It is important to determine this interrelationship because the nonrealisation of a benefit might have a negative impact on another related benefit.
  - *Impact of the benefit:* The impact of the benefit needs to be quantified, and even intangible benefits should in some way or other be quantified. The impact of the benefit relates directly to the sustainment of the benefits.
  - *Classification:* Benefits can be classified into various categories, such as sales, profitability, or productivity. Classifications are used to determine where benefits will have the greatest impact when they are all realised.
- ***Dependencies:*** Benefits depend on various processes and other activities within the organisation. The purpose of this dimension is to have an informed view of the dependencies that will have either a positive or negative impact on the realisation of the benefit.
  - *The business changes on which the benefit depends:* The delivery and sustainment of benefits are dependent on organisational changes. Some of these organisational changes include changing the way people work or even changes to existing processes. These business changes need to be identified as early as possible to ensure that they actually occurred. If these changes did not occur, it might have a negative impact on the full realisation of the benefits.
  - *Earlier benefits on which the benefit depends:* Just as a benefit contributes to other benefits, other benefits might contribute to a specific benefit. These earlier benefits that a benefit is dependent on must be identified. The nonrealisation of earlier benefits will have a negative impact on the realisation of a benefit.

- *Risk of nonachievement of benefit:* The business case normally associates the benefits of a deliverable with success. Organisational leaders must ask the question, What are the risks when some benefits are not realised and the organisation does not actually receive what was promised? This forms part of the larger risk assessment of the programme or project.
- ***Benefit tracking information:*** Benefits need to be tracked during the entire BRM process. The following information makes it easier for all the role players to track the benefits:
  - *The person accountable for the benefit:* Who is accountable for realising the benefit needs to be stipulated. This can be determined from the roles and responsibilities discussed in the previous section.
  - *The person(s) who will receive the benefit:* This information is required to determine who will receive the benefit of the deliverable. It does not speak to who the owner is of the benefit but who will be benefiting from using the deliverable.
  - *Measurement information:* Benefits needs to be measured to determine if they actually deliver on what was promised in the business case. Some information that should be captured include (i) how the measure will be tracked; (ii) what the baseline and target values comprise; (iii) when the expected delivery date of the benefit will be; (iv) when it will be fully realised and sustained; and (v) how frequent reporting will occur.

Knowing the dimensions of a benefit makes it easier for all the role players to understand what the specific benefit is all about, what the interdependencies are, and how the role players will track the benefit. The information captured by the various dimensions can be used to visually display the different benefits in relation to the organisational strategies and objectives. This is typically done in the form of a benefits dependency network.

8.6 Benefits Dependency Network (BDN)

Ward and Daniel (2012) are of the opinion that the BDN is the how of implementing benefits. A BDN is a structured diagram that visualises multiple cause–effect relationships between enablers, changes, and benefits. A BDN template is illustrated in Figure 8-3. The BDN is constructed from the profile of a benefit.

The entire BDN is dependent on the business objectives, and when and if business objectives change, the BDN will change as well. In most instances, a one-to-many relationship exists between the business objectives and the benefits that need to be realised.

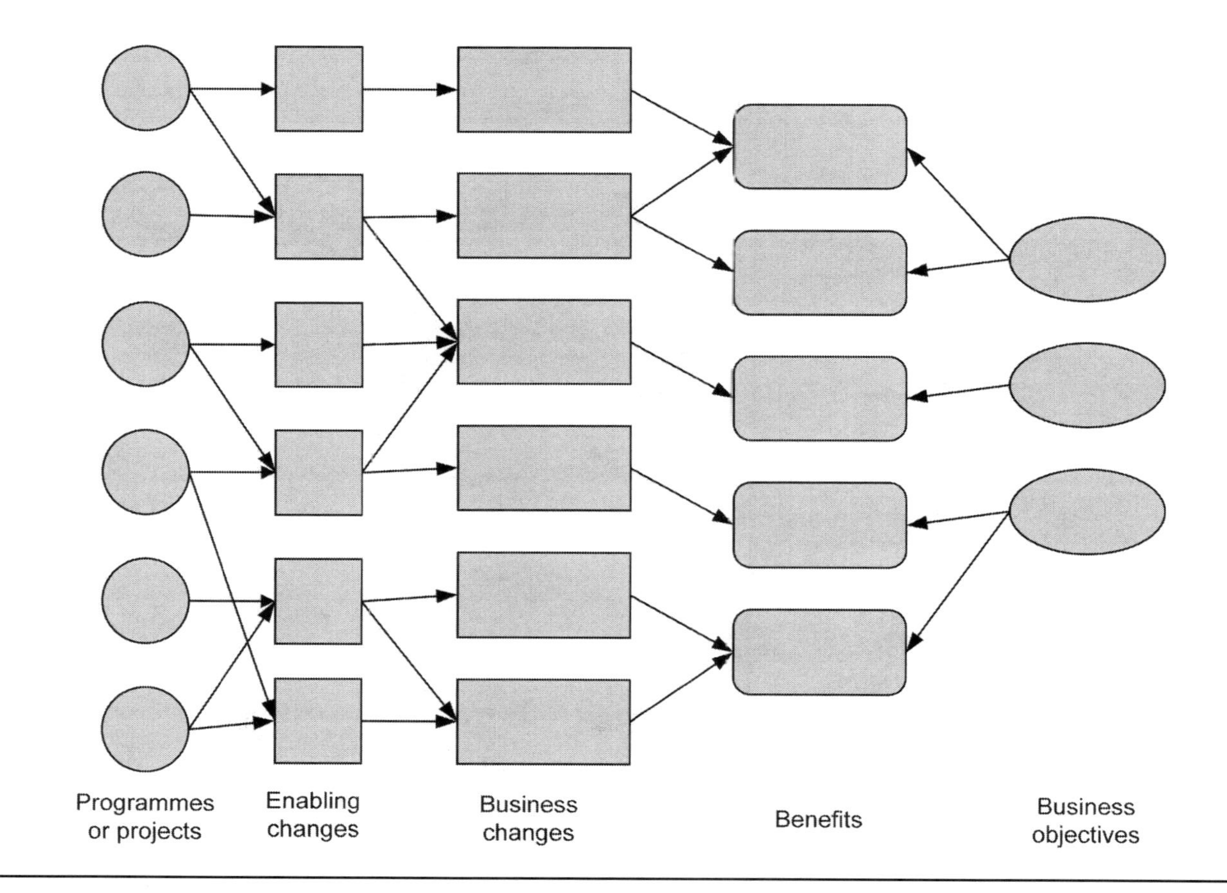

Figure 8-3 Benefits Dependency Network (BDN)

Table 8-1 Benefits Tracking and Monitoring Template

#	Benefit Description	Measure	Baseline Value	Target Value	Start Period	End Period	Reporting Period 1	Reporting Period 2	Reporting Period 3	Reporting Period N	Person Accountable
1	Up-to-date applications for customers	Increase in number of deployments	200	250	January 2018	July 2019	March 2018	June 2018	September 2018	July 2019	Ginger Levin

On the left of the BDN you have the normal programmes or projects that will deliver benefits that contribute to the business objectives. Two important aspects of this are the enabling and business changes. These are the changes that are necessary for the benefits to be delivered and sustained. Business changes are new or different ways of working that are a future and permanent requirement to sustain benefits. Enabling changes are once-off changes that are necessary to allow the business changes to be realised (Ward & Daniel, 2012).

The BDN clearly highlights the interdependencies and provides a visual presentation to the organisational leaders on how projects, organisational changes, benefits, and business objectives are linked and related to one another. The BDN, together with the BRP, should be used to identify, manage, deliver, and sustain the benefits.

Two important aspects are highlighted by the BDN. The first shows how benefits are tracked and monitored throughout the BRP, and the second addresses the notion of change management. These two aspects are discussed in the following sections.

8.6.1 Tracking and Monitoring Benefits

Aguilera (2016) mentions that tracking and monitoring benefits is actually a decision-making tool for organisational leaders. The metrics resulting from tracking and monitoring are used to influence decisions based on project alignment and, ultimately, benefit sustainment. The benefit profile plays an important role in tracking and monitoring benefits. The third dimension of the benefit profile provides project managers with the relevant information to track benefits. Every aspect of the benefit needs to be tracked and monitored. It is therefore advisable that a reporting mechanism is created. This might be as simple as a spreadsheet in which all the relevant information is portrayed, as per Table 8-1.

It is all well and good to track and monitor the benefit, but what happens with the information? The information needs to be shared among all the role players to promote transparency as well as to flag and highlight potential risks that might have an impact on the realisation of the benefits.

8.6.2 Change Management

Benefits cannot be realised without substantial changes within the organisation. As per Figure 8-3, two types of changes are required for the successful delivery of benefits. The most important change focuses on the business changes, wherein the focus is on how the business at large must change to (i) receive the benefits; and (ii) sustain the benefits.

Communication is essential among all the role players to ensure that the appropriate changes are made. It is the responsibility of the benefits or business owner to ensure that these changes occur. Changes might include new or altered processes, new ways of working, and even a change in the business culture. Regardless of the change, each change should be implemented in full to realise the maximum benefit.

8.7 Benefits Realisation Management Maturity

The ultimate goal of BRM is to deliver on all the identified benefits and ensure that these benefits are sustained. This can only be accomplished when there is a level of BRM maturity within the organisation. The Boston Consulting Group (2016) highlighted that organisations that are BRM mature are 1.6 times more likely to realise the identified benefits and up to three times more likely to achieve return on investment (ROI) on individual projects.

Unlike normal maturity models, BRM maturity does not go through phases or levels, but rather focus on building blocks that need to be in place. Three building blocks have been identified thus far:

1. ***Managing the project portfolio:*** Chapter 6 highlighted the notion that the success of a project is not determined on the triple constraint but rather on a continuum. The achievement of business and strategic success is the ultimate level of success. This is typically measured in terms of the value that benefits realised. Portfolios should take benefits into consideration during the identification of portfolio components. Benefits realisation should become an integral part of portfolio management.
2. ***Creating a dedicated space for dialogue:*** Communication is one of the knowledge areas within the *PMBOK® Guide* (PMI, 2017), and it indicates the importance of communication within the P3 disciplines. Within the realm of BRM, communication or meaningful dialogue should take place between all four of the major role players. Results indicate that the involvement of benefits or business owners, project managers, and executives contribute to 64% of the factors leading to successful benefits realisation. This implies that continuous dialogue must take place between these three role players to ensure that the strategies (executives) are translated into projects (business owners) and benefits are implemented and realised (project managers). This continuous dialogue

identifies potential risks and issues that might have an impact on the full realisation of the benefits.

3. ***Establishing the right conditions for success:*** Organisational leaders should ensure that the right conditions are in place to optimally realise benefits. Emphasis should be given to the following conditions. The first is the behaviour of all the role players. The focus should be on the frequent assessment of projects, making sure that (i) they will still deliver benefits; and (ii) the projects are still aligned with organisational strategies. Second, project managers must have the right skills—that is, technical skills are just not enough anymore. Project managers need to be skilled in leadership as well as strategic management. This allows them to see the bigger picture, as well as the role that the project fulfils within the larger realm of the organisation.

Chasing for a mythical maturity level five is not the goal when it comes to BRM maturity. The goal is to ensure that organisational leaders display a level of maturity that is conducive to the delivery and sustainment of benefits.

8.8 The Executive's Role in Realising Benefits

The realisation of benefits plays such an important role that executive management cannot ignore it. More importantly, executives should be more involved in the management of benefits realisation than with project management. Executives are involved in the justification of a project through the business case, which promises benefits to the organisation at large. It is evident from Figure 8-2 where the sponsor and business owner are involved throughout the BRM process. Executives can engage in this process through the responsibilities delineated in the following sections.

8.8.1 Criteria for Benefits Analysis and Grouping

Problems downstream are usually caused by some a lack of planning and commitment upstream. This is especially the case with BRM. It is of utmost importance that benefits are properly identified and mapped to the relevant business objectives. This identification and mapping cannot occur if there is no clear guidance on how benefits should be identified and classified. This can only be done if there is a formal process in place to identify benefits. Executives should play their part in making sure that a well thought-through benefit profile exists for each and every benefit and that a BDN is constructed for each project. This ensures that benefits are properly identified right at the beginning of the project and thus makes it easier to manage the benefits through delivery and sustainment.

8.8.2 BRM Responsibility and Accountability

Executives must make sure that the roles and responsibilities are clearly defined within the BRM process. Unlike a project in which the project manager and the project team assume most of the responsibilities, the BRM process involves the business from the onset of the project. Who is responsible for benefits identification, benefits management, benefits delivery, and, ultimately, benefits sustainment must be clearly defined. Everyone plays a role in the BRM process, but ultimate accountability can only be with one individual, and the executives must ensure that benefits realisation is linked to a specific role. Just as a new CEO of an organisation must take ownership of his or her predecessor's decisions, so should a new business owner take accountability for benefit sustainment. In other words, the responsibility and accountability of benefits realisation does not lie with an individual, it lies with a specific role within the organisation. It is the incumbent of that role that must make sure that benefits are realised.

8.8.3 Integrating the BRP into the Project Plan

Are the benefits realisation plan and the project plan two separate plans? This is one of the concerns to address. The impression is created that benefits realisation is managed outside a project. This is not the case, and BRM should be part and parcel of any project. It begs therefore that the BRP is incorporated into the project plan. The implication is that the project manager must manage the benefits through the project life cycle, as per Figure 8-2. A further implication is that the merged plan ceases when the project is completed. The business owner must then take ownership of the delivered benefits and determine how the benefits will be sustained during normal operations. What is evident is that there cannot be two separate plans.

8.8.4 Change in Culture

Incorporating BRM into an organisation is a culture change. Executives must make sure not only that they initiate this new culture but also that this culture is then imbedded into the organisation. Without a culture of BRM, the optimum delivery and sustainment of benefits will not be achieved. Executives need to actively support BRM, and more importantly, be more transparent in their decision making.

8.8.5 Take a Strategic View of Benefits

Cooke-Davies (2016) states that benefits should be the centre of portfolio, programme, and project management. The reason is that benefits are the underlying success factor for strategic success. Projects deliver benefits, which are used to determine whether value was created. When benefits are viewed in departmental silos, then the full potential of benefits is not optimally realised. Executives must change the culture where benefits are realised in silos. When a strategic or holistic view is taken with regard to BRM, then the potential for optimum benefits realisation increases. BRM begs to be part of portfolio management, where it can be used to optimise the portfolio.

8.9 Conclusion

This chapter highlights the importance of benefits realisation management (BRM) within the disciplines of portfolio, programme, and project management. The chapter starts with a view of BRM in general, what the challenges are to successfully implement BRM, as well as the process of BRM. The roles and responsibilities are highlighted because there is confusion around who must do what. Within the benefits realisation process, certain individuals have a clear role to fulfil. The benefits profile and associated benefits dependency network (BDN) are discussed, indicating the information that is needed for each identified benefit. Three building blocks are identified that can assist organisational leaders in achieving a more mature state of BRM. The chapter concludes with an overview of the executive's role in BRM.

This chapter highlights the importance of BRM within organisations. BRM is, at the end of the day, the vehicle that delivers value. Through value creation, the strategies and business objectives are realised. As BRM is important, it cannot be managed as a separate process and should be incorporated into portfolio, programme, and project management. Benefits are used to determine the project portfolio, and the benefits realization plan (BRP) is incorporated into the project plan. BRM should become part and parcel of project management, and a culture change is needed to give BRM a rightful place when it comes to managing programmes and projects. What the chapter highlights is that BRM is not an option but that it is actually compulsory.

The project life cycle has changed. Executives need to realise that the project life cycle extends into benefits delivery and benefits sustainment. This, in itself, poses challenges to the way projects are managed as the life cycle extends into operations. The positive side is that projects are therefore integrated into the entire organisational system and cannot be divorced from operations.

8.10 References

Aguilera, I. (2016). *Delivering Value: Focus on Benefits During Project Execution.* Newton Square, PA, USA: Project Management Institute. Retrieved from http://www.pmi.org/-/media/pmi/documents/public/pdf/learning/thought-leadership/pulse/benefits-focus-during-project-execution.pdf

Badewi, A. (2016). The Impact of Project Management (PM) and Benefits Management (BM) Practices on Project Success: Towards Developing a Project Benefits Governance Framework. *International Journal of Project Management, 34*(4), 761–778.

Bennington, P., & Baccarini, D. (2004). Project Benefits Management in IT Projects—An Australian Perspective. *Project Management Journal, 35*(2), 20–31.

Bradley, B. (2010). *Benefit Realisation Management: A Practical Guide to Achieving Benefits through Change* (2 ed.). Surrey, England: Gower Publishing Limited.

Breese, R. (2012). Benefits Realisation Management: Panacea or False Dawn? *International Journal of Project Management, 30*(3), 341–351.

Chih, Y.-Y., & Zwikael, O. (2015). Project Benefit Management: A Conceptual Framework of Target Benefit Formulation. *International Journal of Project Management, 33*(2), 352–362.

Cooke-Davies, T. (2016). *The Strategic Impact of Projects—Identify Benefits to Drive Business Results.* Retrieved from http://www.pmi.org/learning/thought-leadership/pulse/identify-benefits-business-results

Dhillon, G. (2005). Gaining Benefits from IS/IT Implementation: Interpretations from Case Studies. *International Journal of Information Management, 25*(6), 502–515.

Marnewick, C. (2014). The Business Case: The Missing Link between Information Technology Benefits and Organisational Strategies. *Acta Commercii, 14*(1), 1–11.

Marnewick, C. (2016). Benefits of Information System Projects: The Tale of Two Countries. *International Journal of Project Management, 34*(4), 748–760.

Paivarinta, T., Dertz, W., & Flak, L. S. (2007, January). *Issues of Adopting Benefits Management Practices of IT Investments in Municipalities: A Delphi Study in Norway.* Paper presented at the 40th Annual Hawaii International Conference on System Sciences, 2007 (HICSS 2007), Mānoa, Hawaii, USA.

Phillips, R. (2016). *Beyond the Project: Sustain Benefits to Optimize Business Value.* Newton Square, PA, USA: Project Management Institute. Retrieved from http://www.pmi.org/learning/thought-leadership/pulse/sustain-project-benefits-optimize-value

Project Management Institute. (2013). *The Standard for Program Management* (3 ed.). Newtown Square, PA, USA: Project Management Institute.

Project Management Institute. (2016). *Benefits Realization Management Framework.* Newton Square, PA, USA: Project Management Institute. Retrieved from http://www.pmi.org/learning/thought-leadership/series/benefits-realization/benefits-realization-management-framework

Project Management Institute. (2017). *A Guide to the Project Management Body of Knowledge (PMBOK® Guide)* (6 ed.). Newtown Square, PA, USA: Project Management Institute.

Sowden, R. (2011). *Managing Successful Programmes*. London, UK: The Stationery Office.

The Boston Consulting Group. (2016). *Connecting Business Strategy and Project Management*. Newton Square, PA, USA: Project Management Institute. Retrieved from http://www.pmi.org/learning/thought-leadership/series/benefits-realization/connecting-business-strategy-project-management

Ward, J., & Daniel, E. (2008). *Benefits Management: Delivering Value from IS & IT Investments*. West Sussex, England: John Wiley & Sons.

Ward, J., & Daniel, E. (2012). *Benefits Management: How to Increase the Business Value of Your IT Projects* (2 ed.). West Sussex, England: John Wiley & Sons.

Chapter 9

Sustainability in Project Management

~ The great challenge of the twenty-first century is to raise people everywhere to a decent standard of living while preserving as much of the rest of life as possible. ~

— Edward O. Wilson*

Sustainability, sustainable development, global warming, and corporate social responsibility. These are terms that are flying around boardrooms all over the world. Yet there is no general consensus about what they mean, what the impact is of sustainability on the organisation and, for that matter, project management, and where it fits in the whole debate over global warming. Regardless, projects are initiated and implemented within organisations where sustainability is on the corporate agenda. This implies that project managers should also have sustainability on their agendas. The major question is whether sustainability is one of the main agenda topics or is just one of the addendum items, for noting.

To determine whether sustainability should be a main agenda topic or an addendum, this chapter focuses first on the notion of sustainability development, sustainability, and all associated definitions. Various standards and guidelines are also discussed. The second and third sections focus in-depth on incorporating sustainability into project management, and in doing so, ask the question whether P3 managers need new competencies with regard to sustainability. The fourth section addresses the notion of maturity and that project manager should

* Retrieved from https://www.goodreads.com/author/quotes/31624.Edward_O_Wilson

be passionate about sustainability and not just complying with bare minimum legislation. This chapter concludes with a presentation of a sustainability model and the role that executives play in creating a project management sustainability way of working.

9.1 Defining Sustainability

Various concepts and definitions are used interchangeably when sustainability is discussed. Sustainability development (SD) focuses on the Brundtland definition, which states that SD is "development that meets the needs of the present without compromising the ability of future generations to meet their own needs" (World Commission on Environment and Development [WCED], 1987, p. 41). Sustainability itself focuses on three dimensions—economic (profit), environmental (planet), and social (people) (Silvius et al., 2012). This implies that when organisational leaders want to implement any change in the organisation, this change must consider these three dimensions within the definition of SD. Sustainability and SD must not be confused with corporate social responsibility (CSR). CSR refers to business practices involving initiatives that benefit society (Marcelino-Sádaba, González-Jaen, & Pérez-Ezcurdia, 2015). Some CSR practices can be perceived as a public relations exercise when organisational leaders support a charity. On the other hand, when organisational leaders really focus on reducing gas emissions at a factory that are close to a town, then it is also perceived as CSR.

Defining sustainability is easy, but incorporating sustainability into the normal day-to-day running of an organisation is not that easy. Benn, Edwards, and Angus-Leppan (2013) are of the opinion that organisational leaders battle to understand how to address this phenomenon. The opposite is also true when organisational leaders actually do integrate sustainability into all three management levels of the organisation. The success or failure of sustainability is highly dependent on whether organisational leaders understand sustainability and whether they embrace it (Thomas & Lamm, 2012). Sustainability cannot be addressed within an organisation when the silo mentality still exists. For sustainability to succeed, a holistic and integrative approach is required (Benn et al., 2013).

The three sustainability dimensions need to be in balance in order for organisational leaders to deliver on sustainable development. The challenge for organisational leaders is to incorporate sustainable principles not just at a strategic level but also at an operational level. This is done through the alignment of business and project activities with the principles of sustainable development (Hope & Moehler, 2014).

9.1.1 Standards and Guidelines for Sustainability

Over the last couple of years, various organisations have published some guidelines that organisational leaders can use to address sustainability. Some provide principles or concepts that need to be addressed, wheras others provide guidelines on how to report and incorporate sustainability.

1. ***United Nations Global Compact:*** The United Nations identified 10 principles within 4 groupings (United Nations Global Compact, 2014). Two principles are grouped under ***human rights***. Principle 1 states that organisations should support and respect the protection of internationally proclaimed human rights. Principle 2 states that organisations must ensure that they are not complicit in human rights abuses. The second grouping focuses on labour, with four principles. Principle 3 focuses on association, where organisations should uphold the freedom of association and the effective recognition of the right to collective bargaining. Principle 4 states that all forms of forced and compulsory labour must be eliminated. Principle 5 focuses on the effective abolition of child labour, whereas Principle 6 states that discrimination in regard to employment and occupation must be eliminated. The third grouping focuses on the ***environment***, with three principles. Principle 7 addresses the notion that businesses should support a precautionary approach to environmental challenges. Principle 8 focuses on the initiatives that organisations can undertake to promote greater environmental responsibility, and Principle 9 encourages the development and diffusion of environmentally friendly technologies. The last principle and fourth grouping (***anti-corruption***) focuses on the fact that businesses should work against all forms of corruption, including extortion and bribery.
2. ***Global Reporting Initiative (GRI):*** The GRI helps organisations and governments understand and communicate the impact of their business on critical sustainability issues—for example, climate change, human rights, and corruption (Global Reporting Initiative, 2011). The GRI works in cooperation with the United Nations Global Compact. The GRI Reporting Framework contains content that is specific to certain industries but also includes content that is generally applicable. When organisational leaders report on SD, they need to report on the following performance indicators. There are nine ***economic performance indicators*** grouped into three aspects—economic performance, market presence, and indirect economic impacts. There are 30 ***environmental performance indicators***, which are divided into (i) materials; (ii) energy; (iii) water; (iv) biodiversity; (v) emissions, effluents, and waste; (vi) products

and services; (vii) compliance; (viii) transport; and (ix) overall performance. The overall ***social performance indicators*** are divided into six aspects—employment, labour/management relations, occupational health and safety, training and education, and diversity and equal opportunity, as well as equal remuneration for women and men. Nine aspects are reported for the ***human rights performance indicators***: (i) investment and procurement practices; (ii) non-discrimination; (iii) freedom of association and collective bargaining; (iv) child labour; (v) forced and compulsory labour; (vi) security practices; (vii) indigenous rights; (viii) assessment; and (ix) remediation. The ***society performance indicators*** consist of five aspects, which focus on local communities, corruption, public policy, anti-competitive behaviour, and compliance. ***Product responsibility performance indicators*** comprise the final performance indicator, which focuses on (i) customer health and safety; (ii) product and service labeling; (iii) marketing communications; (iv) customer privacy; and (v) compliance.

3. ***ISO 26000/14001:*** ISO 26000 Guidance on Social Responsibility focuses on the contribution of CSR to sustainable development (International Organization for Standardization, 2010). This standard consists of seven clauses. Clauses one and two provide the scope, terms, and conditions. Clause three describes the important factors and conditions that influence CSR as well as what it means and how it should be applied by organisational leaders. Clause four explains the principles of CSR. Clause five focuses on how organisational leaders should recognise CSR within the organisation and in subsequent engagement with the relevant stakeholders. Clause six explains the core subjects and associated issues relating to CSR. The core subjects include topics such as human rights, labour practices, the environment, and fair operating practices. Clause seven provides guidance on the integration of CSR into the organisation. ISO 14001 is an international standard that sets out the requirements for an environmental management system. It helps organisations improve their environmental performance through more efficient use of resources and reduction of waste, gaining a competitive advantage, and the trust of stakeholders.

9.2 Sustainability in Project Management

Sustainability in project management should focus on incorporating the three dimensions (economic, social, and environmental) into the project life cycle. Sustainability should thus be integrated from the initiation phase right through to benefits realisation and sustainment. Just as benefits realization should be integrated into the project life cycle, so should sustainability be incorporated into the disciplines of portfolio, programme, and project management. But apart from this, the product itself should also conform to the product responsibility performance indicators and contribute to the overall sustainability of the organisation.

Despite the various international standards and guidelines, current project management standards fail to address sustainability issues (Hope & Moehler, 2014). The *PMBOK® Guide* briefly mentions sustainability as a requirement that needs to be considered but does not add or provide any additional information on how to incorporate sustainability (Project Management Institute, 2013). Other standards and methodologies, such as the *APMBOK* and PRINCE2®, are also silent when it comes to incorporating sustainability into project management.

An analysis by Garies, Huemann, and Martinuzzi (2013) also indicates that sustainability is not explicitly addressed in the project management standards. Underlying values relating to sustainability are found in the various project management standards. Some of these values include ethics, social sensitivity, transparency, social responsibility, and honesty, as well as fairness. With the lack of guidance from project management standards, Silvius et al. (2012) devised a checklist that project managers can use to measure compliance within their project environment. This checklist is based on the three sustainability dimensions and is illustrated in Table 9-1.

Table 9-1 Sustainability in Project Management Checklist

Economic Dimension	Social Dimension	Environmental Dimension
ROI Business agility	Society and customers Labour practices and decent work Ethical behaviour Human rights	Water Materials and resources Transport Energy Waste

Note: Data derived from Silvius et al., 2012.

Each of the 11 checklist items has sub-items that need to be incorporated and checked. There are 36 items that need to be checked and incorporated into the project life cycle. The checklist itself does not guarantee that sustainability will be incorporated into project management. GPM® Global (2014) is of the opinion that the project's product or deliverable should be the focal point when it comes to sustainability in project management. The product itself should adhere to the sustainability dimensions, but these dimensions should also be incorporated into the project life cycle, processes, and even the various knowledge areas.

Project managers need to balance the three dimensions against the needs of the project and that of the larger organisation. This fine balancing act is complicated by two additional considerations, as per Garies et al. (2013). The first consideration is that the short-term goals of a project are in direct conflict with the long-term goal that the organisation must be sustainable. Project managers must make decisions based on what is best for the project while also taking into consideration what is best for the organisation in the long term. The second complication is that decisions and actions taken by the project manager have a direct impact on the local, regional, and global scene. Decisions and their consequences have a more far-reaching impact, and what seems to be a good decision for the project might have a negative impact on some community located at the other end of the world.

Reporting on sustainability also becomes an important aspect that project managers must consider. Sustainability reporting can become part of the normal status reports, but a comprehensive report needs to be submitted at the end of the project. This report can follow the guidelines of the GRI. The individual sustainability in project management reports can then be consolidated at a portfolio management level into a comprehensive report highlighting how the portfolio itself contributes to the organisation's sustainability agenda. Incorporating and reporting on sustainability creates additional work for project managers. However, Deland (2009) states that there are some benefits derived by the project manager from thinking about sustainability. Some of these benefits are that value is added to the project manager's role, leadership and negotiation skills are developed, the visibility of the project manager is increased, and most importantly, it assists the project manager with new ways of thinking—for example, systems thinking and lean practices.

Sustainability in project management, similar to benefits realisation management (BRM), has an impact on the way that the portfolio and subsequent programmes and projects are managed. Sustainability in project management should be viewed in the same way that benefits are viewed, and a tool such as the benefits dependency network (BDN) can be used to create a sustainability impact network (SIN). Such a tool can provide all of the stakeholders with a comprehensive view of sustainability in the portfolio as well as how the various programmes and projects have an impact on each other's sustainability agenda.

9.3 Sustainability in Project Management Competencies

Understanding sustainability and having models that guide project managers on how to incorporate the sustainability dimensions are of no value when the project manager does not have the required competencies to incorporate sustainability. Silvius and Schipper (2014) opine that it is the responsibility of project managers to realise sustainability, and therefore they should be competent in sustainable development. It must be noted that none of the competency frameworks explicitly address the notion of sustainability. The IPMA ICB® mentions sustainability as one of many competencies whereby project managers can ensure that the project contributes to the mission and the sustainability of the organisation (International Project Management Association [IPMA], 2015). Sustainability is also addressed in the IPMA ICB® within the competencies of (i) compliance, standards and regulations; (ii) culture and values; (iii) personal integrity and reliability; and (iv) procurement.

Project managers should have the skills, competencies, and knowledge to change behaviour with regard to the three sustainability dimensions without "these changes merely being a reaction to existing problems" (de Haan, 2010, p. 320). In this publication, de Haan identifies 12 competencies that can assist project managers in changing their behaviour, as follows:

1. Knowledge should be gathered with an openness to a world in which new perspectives are continuously integrated.
2. Project managers should think and act in a forward-looking manner.
3. Knowledge should be acquired in an interdisciplinary manner.
4. Project managers should be able to deal with incomplete and overly complex information.
5. Project managers should cooperate in decision-making processes where different points of view concerning sustainable development are analysed, and where controversies are dealt with in a discursive manner.
6. Project managers should be able to cope with individual problematic situations in which the impact of current actions is assessed as well as how the impact of current actions affects their own future of decision making.
7. Project managers must participate in a collective decision-making processes.
8. The project manager should be able to be an activist for sustainability and motivate the team members to also become activists.
9. Reflection is an important competence whereby the project manager should reflect upon his or her own principles as well as those of the team members.
10. Equity in decision making and planning should be embraced and promoted.
11. The project manager should plan and act autonomously and should know and reconsider his or her own personal rights, needs, and interests to achieve sustainable development.
12. The project manager should show empathy for and solidarity with the disadvantaged.

This list of competencies speaks directly to the core of who and what the project manager is and not necessarily to hard competencies, as described by the various competency frameworks. Sustainable development competencies become a way of life and are assimilated into the way the project manager works.

9.4 Maturity Model for Sustainability in Project Management

The notion of maturity is prevalent within the project management discipline. The same applies to the maturity of sustainability within project management. The maturity of sustainability in project management can be measured at two levels. The first level, as identified by Silvius and Schipper (2015), focuses on the maturity of sustainability in project management, whereas the second level focuses more on the deliverable of the project and how the deliverable contributes to sustainability development (Hope & Moehler, 2014).

Project managers' commitment to sustainability is determined at five levels or stages. The levels move from reactive to proactive and illustrate the project managers' commitment to sustainability principles. The first level is when project managers fail to even comply with prevailing regulations. The project managers are opportunistic and do not engage with the notion of sustainability. Level two is achieved when the project manager complies with all the relevant environmental and social regulations. This level can be compared with level three of the project management maturity models (PMMM), where the basic processes are in place. Level three focuses on how project managers introduce sustainability activities. It must be noted that these activities are based on individual project manager's efforts, and that it is not a holistic approach that is embraced by all project managers. Level four is reached when everyone in

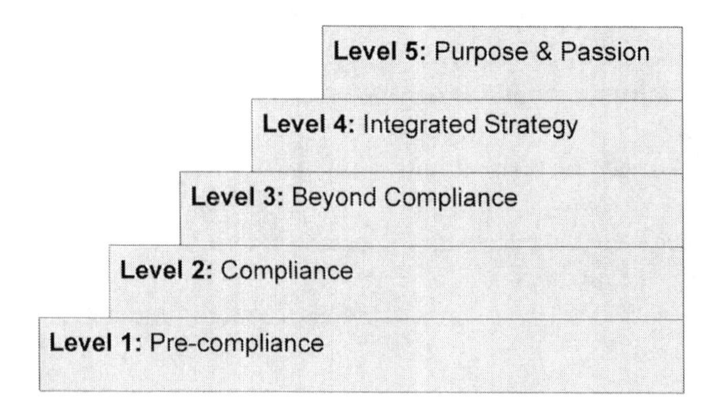

Figure 9-1 Maturity Levels of Sustainability in Project Management (Adapted from Silvius and Schipper [2015])

the project management discipline (portfolio managers, programme managers, project managers, and the project management office [PMO]) realises the importance of sustainability and the added value that they can gain from sustainable activities and from integrating sustainability into the various projects' strategies. Level five is attained when projects also create a sense of responsibility to improve society and the environment and contribute to a better world (Silvius & Schipper, 2015). These five levels are illustrated in Figure 9-1.

Hope and Moehler (2014) are of the opinion that projects themselves can be categorised based on the level of sustainability composition. Four categories are identified, as follows: (i) sustainable by definition; (ii) sustainable by project impact; (iii) sustainable by deliverable impact; and (iv) sustainable in general (Hope & Moehler, 2014, p. 363). There is an overlap with the levels of Silvius and Schipper (2015) but with an additional focus on the impact of the product itself. Although project managers should incorporate sustainability principles into their respective projects, focus should also be on the impact of the deliverable on sustainability development. It must be noted that more time and effort is spent on a project when the project is labeled at a higher level or category.

9.5 Project Management Sustainability Model

The previous section highlighted that to achieve a maturity level five, sustainability principles should be incorporated throughout the entire project life cycle. Regardless of the project management standard or methodology that is followed, sustainabilty, as with benefits, should be one of the concerns of a project manager.

The first concern that needs to be addressed is how to incorporate the sustainability dimensions into the various knowledge areas. The checklist provided by Silvius et al. (2012) can be used. A practical example is the knowledge area of Project Procurement Management. When decisions must be made on the procurement of materials, the following questions should be raised: What is the impact on the environment when the materials are transported? Was child or forced labour used to manufacture the material? What is the impact on the environment during the manufacturing of the specific material? Using the checklist as a basis, an awareness is created among the project team, stakeholders, and sponsors of the sustainability dimensions. This awareness should eventually lead to a point at which maturity level five is reached, where sustainability is the purpose and passion. Each knowledge area can be scrutinised using the checklist as a basis to consider as one works to achieve level five.

Second, to achieve mature project management sustainability levels, the sustainability dimensions should be incorporated into the life cycle and processes of the project. Looking at sustainability from a knowledge area perspective

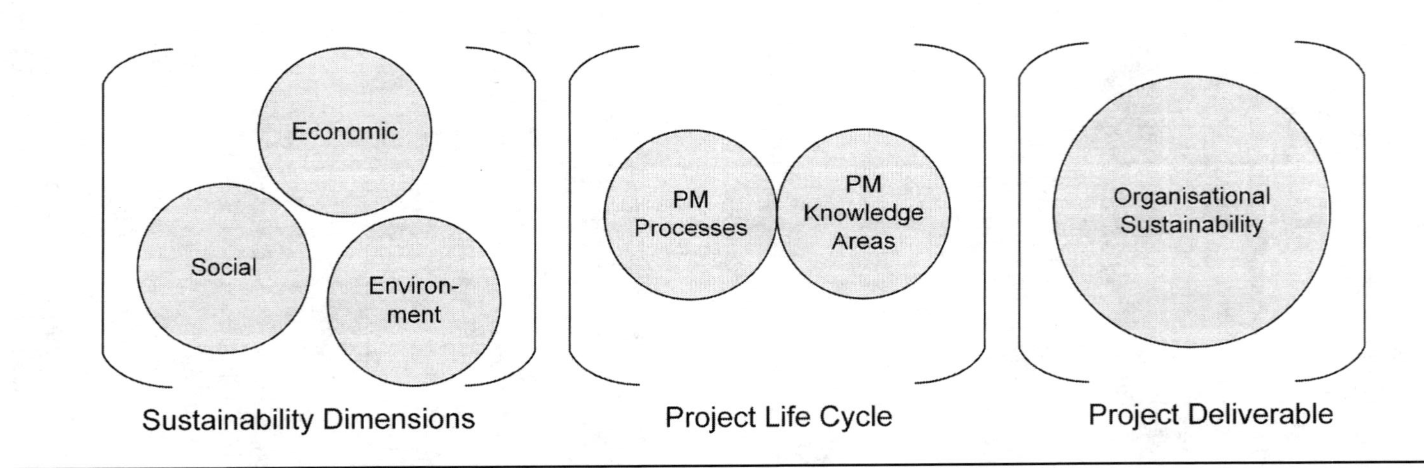

Figure 9-2 Project Management Sustainability Model

creates a once-off view, but when sustainabiltity is viewed as part and parcel of the entire project life cycle, then sustainability is embedded into the entire project. The ultimate goal is to report on sustainability at the end of the project using the GRI as a guideline.

The third concern is whether the deliverable of the project itself contributes to the sustainability of the organisation. A sustainable organisation is achieved when the ultimate benefit and deliverables of the various programmes and projects contribute to the sustainability of the organisation. Figure 9-2 provides a simplified project management sustainability model.

The three sustainability dimensions feed or are incorporated into the project life cycle. The project life cycle consists of the various project management (PM) processes and knowledge areas. The sustainability in project management checklist can be used to determine how sustainability should be incorporated or addressed during the project life cycle. At the end of the project life cycle, a deliverable is introduced into the organisation. This deliverable introduces change and benefits to the organisation, and the ultimate goal should be that the various project deliverables contribute to the sustainability of the organisation.

9.6 The Executive's Role in Sustainability Project Management

A positive relationship exists between effective corporate governance and sustainability development (Aras & Crowther, 2008). In order for sustainability to filter down the organisational hierarchy and find a rightful place in the management of the project portfolio, executives must ensure that they abide and adhere to the principles of corporate governance. Sustainable development should then become a way of working within the organisation. With regard to projects and project management specifically, executives can push the sustainability agenda in various ways.

9.6.1 Promote the Sustainability Agenda

The very first thing that executives should do is to place SD on the organisational agenda. SD should be an agenda item that receives a high level of attention as well as the necessary funding. The principles of SD should be embraced by the executives. They should not merely pay lip service to these principles to keep the environmentalists satisfied. Recycling is an excellent example. Collection holders for recycled paper should be placed all over the organisation, and a life style of recycling should be inculcated. When sustainability is high on the corporate agenda, then it will naturally find its way to the project management agenda.

9.6.2 Integration of Sustainability into P3 Discipline

Just as SD should be an organisational agenda item, it should also be part of the P3 agenda. Just incorporating sustainability into project management is not enough. SD should be part of the entire discipline, and executives should focus on two aspects. The first aspect is that portfolio, programme, and project managers should be knowledgeable about sustainability. This might imply some training or awareness, but the P3 managers cannot be oblivious of sustainability. The second aspect that executives can enforce is to ensure that SD is addressed at all levels of portfolio, programme, and project management. Just as there is a sustainability checklist for project management, there should be a checklist for portfolio and programme management as well. Addressing SD at all three levels ensures that the sustainability agenda is promoted within the discipline as well as within the organisation at large.

9.6.3 Enable P3 for Maturity Improvement

SD should mature from pre-compliance or compliance levels to a maturity level at which everyone involved in projects and project management are passionate about sustainability and the role they can play in promoting SD. Gaps need to be identified, highlighted, and rectified to ensure the attainment of maturity level five. In contrast with project management maturity (PMM), where a maturity level three is good enough, a maturity level four or five is preferred in the case of sustainable project management. Each project is measured based on its sustainability dimensions and to what extent the various aspects have been addressed, given the five maturity levels. Identified areas of non-compliance should be addressed through training, awareness, and, most commonly, through a change of attitude.

9.6.4 Sustainability Reporting

Executives must report on an annual basis on SD. This compliance must be pushed down to programme and project managers. Reporting on sustainability in projects must be compulsory and should be one of the success criteria. The reporting should be based on the GRI guidelines. Sustainability reporting will create an additional workload for project managers. But when executives take SD seriously, then sustainability reporting should be addressed. It must be made clear that sustainability reporting forms part of the larger picture and that reporting is compulsory and not just a nice-to-have.

9.6.5 *Competencies in SD*

The previous chapters highlighted the importance of the various competencies that project managers should master. The competencies highlighted in this chapter differ from the more technical type of competencies that were discussed earlier. Executives should ensure that portfolio, programme, and project managers should gain the necessary competencies in SD. Unfortunately, these competencies are not that easily acquired or mastered. The competencies highlighted by de Haan (2010) are not learned by going to a course and getting certified based on it. These competencies are almost a way of life and speak directly to the nature and soul of the project manager. These competencies bloom under the ideal situation in which sustainability is part of the organisational culture. Where the culture is based on profits only, then these competencies will not flourish. It is thus the duty of the executives to create an environment in which sustainability is cherished. In their special report on sustainability, Kiron et al. (2013) highlighted 10 capabilities that distinguish organisations leading in SD. These capabilities can only be initiated at an executive level, which implies that the sustainability agenda is driven from the top and that executive support is compulsory for sustainability in project management.

9.7 Conclusion

This chapter highlights the notion of sustainability in project management (PM). Although various publications and research studies indicate the importance of sustainability, sustainability itself is not addressed in the current project management standards and methodologies. The first section focuses on general definitions of sustainability and the various standards and guidelines that can assist an organisation to incorporate sustainability into the organisation. The second section focuses specifically on sustainability in project management. The checklist of Silvius et al. (2012) is presented as a way to incorporate sustainability. To incorporate sustainability into project management, project managers must have certain competencies, and this chapter highlights that these competencies cannot be taught but are actually a way of life. The notion of maturity was also addressed, where the idea is to move away from compliance to a stage or level at which sustainability forms part of an integrated strategy. A conceptual model indicates that sustainability should be addressed across the entire value chain and not just within the formal project management process. The project deliverable must be sustainable and contribute to the sustainability of the organisation.

Sustainability is not a choice anymore, and portfolio, programme, and project managers should realise that this is another aspect that needs to be added to the overall complexity of a project. Project managers must familiarise themselves with sustainability and how to incorporate it into the project management life cycle. They should also realise that the sustainability of their projects and project deliverables contribute to the overall sustainability of the organisation. Just as benefits contribute to the overall well-being of the organisation, so does sustainability contribute to the overall well-being of the organisation.

In a world in which normal citizens expect organisations to be responsible corporate citizens, sustainability is perceived as the yardstick to measure this responsibility. Projects *per se* contribute to the organisation's responsible corporate citizenship, but this can only be achieved if projects are managed in a sustainable way.

9.8 References

Aras, G., & Crowther, D. (2008). Governance and Sustainability: An Investigation into the Relationship between Corporate Governance and Corporate Sustainability. *Management Decision, 46*(3), 433–448.

Benn, S., Edwards, M., & Angus-Leppan, T. (2013). Organizational Learning and the Sustainability Community of Practice: The Role of Boundary Objects. *Organization & Environment, 26*(2), 184–202.

de Haan, G. (2010). The Development of ESD-Related Competencies in Supportive Institutional Frameworks. *International Review of Education, 56*(2), 315–328.

Deland, D. (2009). *Sustainability through Project Management and Net Impact.* Paper presented at the 2009 PMI Global Congress, Orlando, FL, USA. http://www.pmi.org/learning/library/sustainability-goals-achieving-framework-technique-6776

Garies, R., Huemann, M., & Martinuzzi, A. (2013). *Project Management and Sustainable Development Principles.* Newtown, PA, USA: Project Management Institute.

Global Reporting Initiative. (2011). *Sustainability Reporting Guidelines.* Retrieved from https://www.globalreporting.org/resourcelibrary/G3.1-Guidelines-Incl-Technical-Protocol.pdf

GPM® Global. (2014). *The GPM® P5™ Standard for Sustainability in Project Management* (p. 30). Noir, MI, USA: GPM® Global.

Hope, A. J., & Moehler, R. (2014). Balancing Projects with Society and the Environment: A Project, Programme and Portfolio Approach. *Procedia—Social and Behavioral Sciences, 119*(0), 358–367.

International Organization for Standarization. (2010). *ISO 26000—Social Responsibility* (p. 118). Geneva: Switzerland: International Organization for Standarization.

International Project Management Association. (2015). *Individual Competence Baseline for Project, Programme & Portfolio Management,* Version 4.0 (p. 416). Zurich, Switzerland: International Project Management Association.

Kiron, D., Kruschwitz, N., Rubel, H., Reeves, M. J., & Fuisz-Kehrbach, S. (2013). *Sustainability's Next Frontier: Walking the Talk on the Sustainability Issues That Matter Most.* Retrieved from http://sloanreview.mit.edu/projects/sustainabilitys-next-frontier/

Marcelino-Sádaba, S., González-Jaen, L. F., & Pérez-Ezcurdia, A. (2015). Using Project Management as a Way to Sustainability. From a Comprehensive Review to a Framework Definition. *Journal of Cleaner Production, 99,* 1–16.

Project Management Institute. (2013). *A Guide to the Project Management Body of Knowledge (PMBOK® Guide)* (5 ed.). Newtown Square, PA, USA: Project Management Institute.

Silvius, A. J. G., & Schipper, R. P. J. (2014). Sustainability in Project Management Competencies: Analyzing the Competence Gap of Project Managers. *Journal of Human Resource and Sustainability Studies, 2*(2), 40–58.

Silvius, A. J. G., & Schipper, R. (2015). Developing a Maturity Model for Assessing Sustainable Project Management. *The Journal of Modern Project Management, 3*(1), 17–27.

Silvius, A. J. G., Schipper, R., Planko, J., Van den Brink, J., & Köhler, A. (2012). *Sustainability in Project Management.* Surrey, England: Gower Publishing.

Thomas, T., & Lamm, E. (2012). Legitimacy and Organizational Sustainability. *Journal of Business Ethics, 110*(2), 191–203.

United Nations Global Compact. (2014). *Guide to Corporate Sustainability.* Retrieved from https://www.unglobalcompact.org/docs/publications/UN_Global_Compact_Guide_to_Corporate_Sustainability.pdf

World Commission on Environment and Development. (1987). *Our Common Future* (Oxford Paperbacks). Oxford, UK: Oxford University Press.

Chapter 10

Project Governance and Auditing

~ The time is always right to do right. ~

— Martin Luther King, Jr.[*]

This chapter focuses on doing the right thing in the project environment within the larger realm of corporate governance. Corporate governance is a topic of discussion after all the various scandals that have plagued the corporate scene. It is therefore logical that this discussion flows over into project governance. This discussion is especially important when large infrastructure projects are implemented, such as the Panama Canal expansion or the Marmaray Tunnel in Turkey. Although the temptations are greater within large infrastructure projects, a culture of governance should be instilled in each and every project, regardless of size and cost. The focus should be to always do right, even with extenuating circumstances. The concepts of benefits realisation management (BRM), sustainability development (SD), and governance are interlinked, and a breakdown in one concept will have a negative impact on the other two concepts.

This chapter is divided into two parts. The first part focuses on project governance, and the second part focuses on project auditing. The relationship between corporate governance and project governance is addressed in the first section. The second section focuses on project governance and its relation to the governance of project management and governmentality. The third and fourth

* Retrieved from https://www.brainyquote.com/quotes/quotes/m/martinluth106169.html

sections focus on project governance and project success rates and the respective roles and responsibilities. The fifth section provides a project governance framework. The sixth section introduces the second part and focuses on project auditing. This is followed by a discussion on the various types of project audits. A model for continuous project assurance is provided in section eight, and the chapter concludes by the positive role that executives can play in the establishment of a project governance and auditing culture.

10.1 Corporate and Project Governance

The need for corporate governance emerged from the expansion of mercantile and business operations, stretching from the 14th century traders to the modern conglomerates of the 21st century (Clarke & Branson, 2012). With the advent of registered public and private organisations, legislation in various countries requires that the financial position and activities of organisations are reported and disclosed to shareholders. This is done by way of implementing formal audit practices. Various definitions and interpretations of corporate governance exist—for example, the World Bank perceives corporate governance as a way to maximise value subject to meeting the organisation's financial and other legal and contractual obligations (Iskander & Chamlou, 2000). Corporate governance is also seen as the establishment of structures and processes with the appropriate checks and balances that enable directors to discharge their legal responsibilities (Institute of Directors Southern Africa, 2016).

To ensure that corporate governance is being implemented, various countries introduced legislative measures or guides for corporate governance. After the corporate collapses of the 1980s, governance codes were established by way of combining corporate legislation, exchange rules, accounting standards, and organisational law. Various perspectives exist that address corporate governance issues. The major reports or perspectives are as follows:

- ***Cadbury Report:*** This report, officially known as *Financial Aspects of Corporate Governance*, provides recommendations on the arrangement of company boards and accounting systems to mitigate risks and failures associated with corporate governance (Cadbury, 1992, 2002).
- ***Organisation for Economic Co-operation and Development (OECD):*** This report focuses on six principles that "are intended to help policy-makers evaluate and improve the legal, regulatory, and institutional framework for corporate governance, with a view to support economic efficiency, sustainable growth and financial stability." (OECD, 2015, p. 9). The six principles focus on (i) the corporate governance framework; (ii) the rights and equitable treatment of shareholders and key ownership

functions; (iii) institutional investors and stock markets; (iv) stakeholders; (v) disclosure and transparency; and (vi) the responsibilities of the board.

- ***Sarbanes–Oxley Act of 2002:*** This act was enacted as a reaction to a number of major corporate and accounting scandals. There are 11 sections that prescribe the responsibilities of the board of directors of private organisations (Moeller, 2008). These 11 sections are (i) the establishment of a Public Company Accounting Oversight Board; (ii) auditor independence; (iii) corporate responsibility; (iv) enhanced financial disclosures; (v) reporting of conflict of interests; (vi) the authority and resources of the Securities and Exchange Commission (SEC); (vii) studies and reports; (viii) corporate and criminal fraud accountability; (ix) the enhancement of white collar crime penalties; (x) corporate tax returns; and (xi) the accountability of corporate fraud (Maurizio, Girolami, & Jones, 2007).
- ***King IV:*** This South African report contains principles as well as recommended practices aimed at strengthening corporate governance (Institute of Directors Southern Africa, 2016). This report focuses on four areas. The first area highlights the governing body's primary governance role and responsibilities. The second area discusses 17 principles that focus on the journey of effective corporate governance. Practices that support and give effect to the principles form the third area. The fourth area covers the actual governance outcomes. These are benefits that organisations could realise through effective governance, including an ethical culture, performance standards, effective control, and legitimacy.

These various corporate governance reports highlight the importance of governance as well as stress the different approaches that countries and regions apply. At the end of the day, it is a balancing act between doing things right, on the one hand, and creating profits and value for the stakeholders, on the other hand.

Project governance should be seen in relation to corporate governance, where corporate governance informs and guides project governance (Müller, 2009; Project Management Institute [PMI], 2016). If project governance is a subset of corporate governance, then the same logic should apply in that project managers should do the right thing within the project while taking the project stakeholders into account. The next section investigates the notion of project governance and the various definitions that are used.

10.2 Defining Project Governance

In spite of research into the phenomenon of project governance, various authors agree that there is still no generally agreed-upon definition for it (Bekker, 2014; Brunet & Aubry, 2016). To make matters even worse, the concepts of project governance, governance of projects, governance of project management, as well as governmentality, are sometimes used interchangeably or in the wrong context. Müller (2009) claims that the aim of project governance should be that all projects are delivered in a consistent and predictable way. Chapter 3 highlights that projects should be aligned to the organisational strategy, and project governance is the vehicle to ensure that this occurs. PMI (2016) provides a broader definition by stating that project governance comprises the framework, functions, and processes that guide the various project management activities. At the end of the day, project managers should deliver a product or service with the knowledge that they always did the right thing to the best of their knowledge. Turner (2006, p. 93) defines project governance as the "structure through which the objectives of the project are set, as well as the means to attain these objectives and to monitor the performance against these objectives." It is evident from these three definitions that there must be a structure in place, and this structure should guide the various project activities in order to deliver a successful project.

Project governance should not be confused with the governance of projects. The governance of projects is at a portfolio or programme management level—focusing on how projects are selected, coordinated, and controlled (Badewi, 2016). The governance of projects refers to portfolio and programme governance.

10.2.1 Governance of Projects versus Project Governance

The governance of projects focuses on the extent to which organisations are run through projects. Müller et al. (2016) reviewed the current literature and stated that the governance of projects is the way that organisational leaders govern groups of projects within the organisations. This therefore implies that the governance of projects occurs at a more strategic level—that is, portfolio and programme management—whereas project governance occurs more at the operational level—that is, project management. Bekker (2015) adds more complexity to the notion of project governance and distinguishes between three levels of project governance. The first level focuses on project governance within a specific organisation, whereas the second level focuses on governance between two or more companies. The third level tends to view projects as temporary organisations that define appropriate governance frameworks within which various project decisions should be made.

10.2.2 Governmentality

Governmentality means managing the perceptions, attitudes, values, and culture needed to govern and direct projects in order to deliver project value (Badewi,

2016). Governmentality is also described as the way to govern (Lemke, 2015). In addition, governmentality focuses on the rationale for project governance as well as the various attitudes that exist towards project governance (Müller, Zhai, & Wang, 2017). The focus is on the attitude that organisational leaders and project managers display toward governance in general and project governance specifically. When project managers have a low governmentality, then the chances are good that project governance will not be adhered to within the project. The ideal situation is one in which project managers portray a high level of governmentality. There are three approaches to governmentality—authoritarian, liberal, and neo-liberal (Müller et al., 2016). When project managers follow the authoritarian governmentality approach, they enforce process compliance and rigid governance structures. The liberal governmentality approach emphasises the control of the project outcomes within a clearly defined but flexible governance structure. The neo-liberal approach ensures that the team members' values are in line with the values of the project. This alignment fosters self-control within a governance structure.

10.3 Project Governance and Project Success

Implementing project governance and all the associated controls and frameworks comes at a cost. It either adds complex layers of approval, which causes delays, or there might be a direct cost involved. The question that often arises is whether adhering to project governance is beneficial to the project and the organisation at large. Apart from the honourable aspect regarding whether organisational leaders and project managers should do the right thing, researchers have found a positive relationship between project governance, governmentality, and project success. Bekker and Steyn (2008) determine that project success improves when there is a governance structure in place and when the organisational leaders adhere to governance principles or governmentality. Joslin and Müller (2016) determine in their studies that project governance is actually the antecedent for project success. Therefore, a direct link between project governance and project success does not necessarily exist, but project governance creates an environment in which it is easier to improve the success of a project. The focus should be on the outcome of the project itself and not necessarily on the controls and processes. At the end of the day, project governance takes a mediating role.

10.4 Project Governance—Roles and Responsibilities

Project governance is advocated when everyone involved in the project understands their respective roles and responsibilities. PMI (2016) is of the opinion that the key roles and responsibilities are that of the governing body, the project sponsor, the project manager itself, the project management office (PMO), as well as other key stakeholders.

10.4.1 Governing Body

The governing body usually consists of executives from the various divisions that are directly involved with projects. These individuals are already familiar with the notion of corporate governance. They are responsible for oversight and ensuring that the project is aligned with the organisational strategies and objectives. These individuals are not directly involved in the day-to-day governance issues but provide guidance on oversight, control, integration, and decision making. The specific responsibilities include but are not limited to establishing project governance policies and processes, ensuring project alignment with the organisational strategies and objectives, and, ultimately, approving the closure or termination of a project. These responsibilities should be in line with the responsibilities and duties of corporate governance.

10.4.2 Project Sponsor

The role of the project sponsor is to provide the necessary resources when and as they are required. The project sponsor is usually a member of the governing body and should actively support project governance. The main responsibilities of the project sponsor can be summarised as follows:

- The sponsor should ensure that there is alignment between the project's strategy and that of the organisation.
- The sponsor should continuously monitor and control the project's deliverable as to ensure project success.
- The project sponsor is also responsible for the removal of any barriers or obstacles that might have a negative impact on project success.

10.4.3 Project Manager

The project manager liaises directly with the governing body and the project sponsor with regard to project governance. The project manager ensures that the governance functions and processes are adhered to as well as implemented. The project manager's responsibility is not just an oversight role but also that of control. The overall role of the project manager is to ensure that there is compliance with the various governance principles. Some of the main responsibilities of the project manager include the following:

- Ensure conformance of the project to governance policies and processes. This implies that the project manager must have knowledge of project as well as corporate governance, thus ensuring alignment between both.
- Act as the principle liaison between the project sponsor, the governing body, and the project team. This liaison focuses on project-related concerns and issues.
- Manage and control the various project management activities and processes in such a way that it ensures project success.

10.4.4 Project Management Office

The role of the PMO was discussed in Chapter 7. The PMO's role in terms of project governance is to ensure adherence to project governance standards and processes and that these are successfully implemented by the project manager. The PMO should support the project manager through the standardisation of the governance process. This standardisation enables project managers to do the right thing in every project and thus improve the governmentality of the project environment.

Project governance is the responsibility of everyone who is involved in or touched by a project. It is not the sole responsibility of the project manager, and everyone must understand his or her specific role and associated responsibilities. When everyone performs his or her roles and responsibilities, then project governance is improved.

10.5 Project Governance Framework

One of the main determinants of project success is an effective project governance structure or framework. Badewi (2016) is of the opinion that an effective project governance framework is a main contributor to project success. A project governance framework is described as an authoritative structure comprising processes and rules that ensure project success (Brunet & Aubry, 2016). Too and Weaver (2014) devised a project governance framework that consists of three levels. The first level is the governance system, which focuses on corporate governance and compliance. The second level focuses on the management system in which the alignment between strategy, corporate governance, and project governance is addressed. The third level is labeled the project delivery system in which the emphasis is on the delivering of the right project within the portfolio. PMI (2016) provides more details outlining what a project governance framework should comprise. According to PMI, a project governance framework consists of four governance domains with specific functions and processes that need to be adhered to during project implementation.

10.5.1 Governance Domains

There are four governance domains within the project governance framework. The first domain focuses on alignment. The focus within this domain is on the internal alignment of project governance. The idea is that the various governance roles oversee the project's alignment with the overall organisational strategies and within the portfolio. The second domain focuses on risk management, and a project should be monitored and audited based on risk exposure. The third domain emphasises the project's overall performance. This domain enables the project manager to collect the necessary performance information and report on a regular basis on the performance of the project. If this is done properly, then there should be no surprises with regard to projects being late or overbudget. The last domain is that of communication. The plan should describe when and how elements of the project should be communicated to the various stakeholders. This governance domain incorporates the project communications management knowledge area (PMI, 2013).

10.5.2 Governance Functions

Within the realm of project governance, certain functions need to be performed on a daily basis. PMI (2016) defines four major functions that need to be performed. These project governance functions are oversight, control, integration, and decision making. Project governance oversight focuses on the provision of guidance, direction, and leadership to the project team. Project governance control concentrates on the day-to-day management of the project and includes the monitoring, measuring, and reporting on project activities. Project governance integration focuses on the entire portfolio and determines how the various programmes and projects are integrated. The decision-making function emphasises the various structures and roles that need to be in place to enhance project governance.

In their research findings, Brunet and Aubry (2016) conclude that an effective project governance framework leads to efficiency within the project, it provides legitimacy to the project, and it creates accountability for all the project stakeholders.

10.6 Contextualising Project Governance

Figure 10-1 provides clarity on how the various concepts are linked together.

A clear distinction should be made between the governance of projects and project governance. Project governance focuses on project management, and the governance of projects focuses on portfolio and programme management. The input into project governance and the governance of project management is governmentality. Without the notion of doing things right, no governance structure is going to work. Governmentality is institutionalised, and the higher the level of governmentality, the better the chances that project governance and the governance of projects will be performed.

Project governance and the governance of projects do not function in isolation. Both project governance and the governance of projects are guided by corporate goverannce and should be governed within the realm of corporate governance. A project manager should therefore align project governance to corporate governance as well as within the broader concept of the governance of projects. Corporate governance, the governance of projects, as well as project governance, are all guided by the vision and strategies of the organisation.

The second part of this chapter focuses on project audits and assurance.

10.7 Project Auditing

The notion of auditing is based on the following principles: (i) a critical examination to establish accuracy, truthfulness, completeness, and compliance; (ii) a systematic process; (iii) the evaluation of evidence assertions about the business actions; (iv) the fair presentation of the evidence that indicates performance; and (v) compliance and communication of the results. Various types of audits exist, and one specific type of audit is the project audit.

According to Reusch (2011), the purpose of a project audit is to determine whether a project meets the strategic requirements. Projects should also be auditable in regard to their compliance with statutory, regulatory, and corporate guidelines. Project auditing can form part of the normal project management process in which the focus is on ensuring that the project team and the project manager have put in place both business and technical processes that are likely to result in a successful project (McDonald, 2002).

10.7.1 Defining Project Auditing

Just as there is no common definition of project governance, is there no common definition of project auditing. PMI itself does not provide a definition for

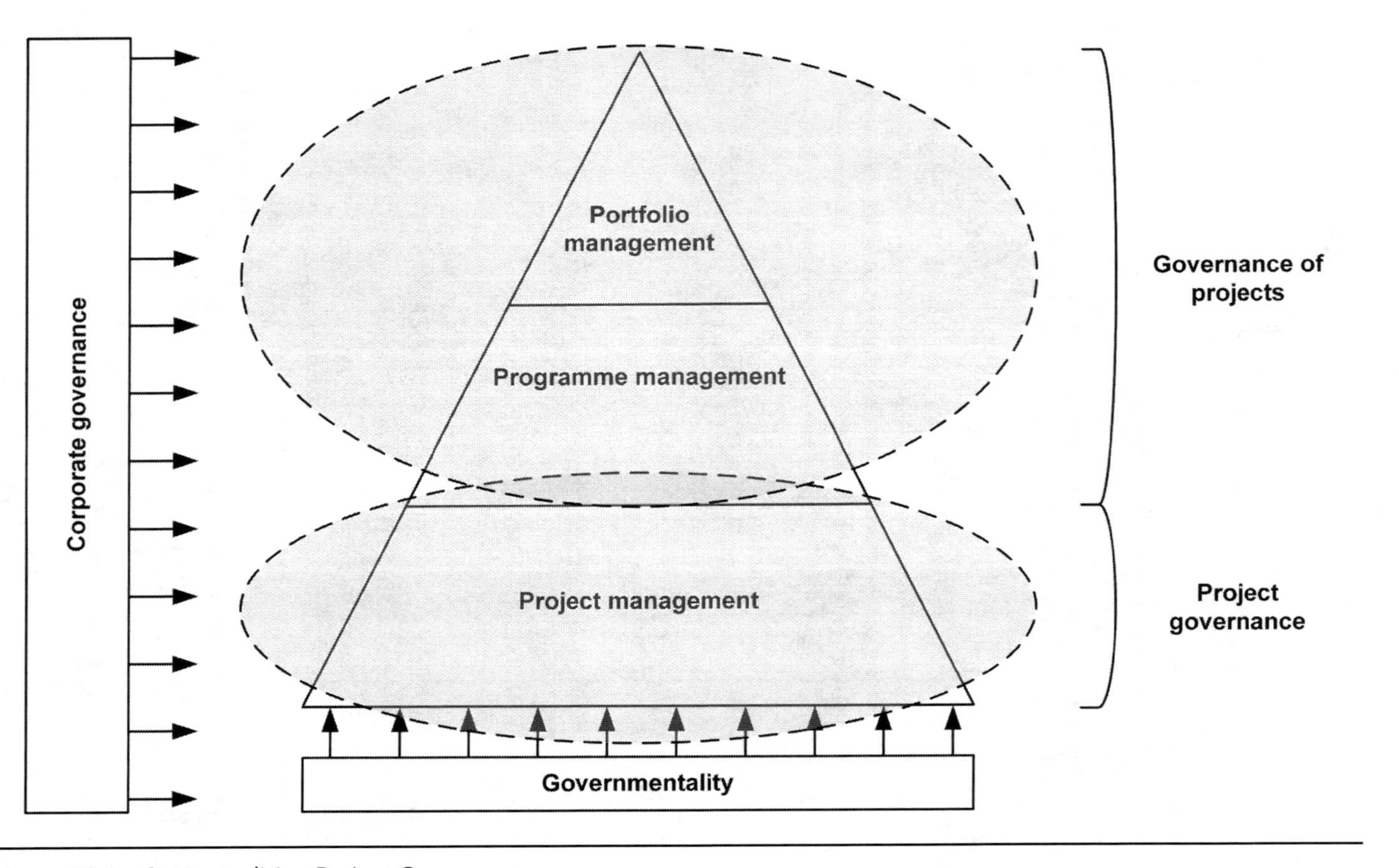

Figure 10-1 Contextualising Project Governance

project auditing, but it does mention that project audits should be performed. Within the *PMBOK® Guide*, assurance is dealt with under the project quality management knowledge area. The emphasis is on whether the appropriate quality standard are used (PMI, 2013). Project assurance in PRINCE2® provides an audit trail that checks whether (i) the project delivers value to the organisation; (ii) the products meet the user's needs; and (iii) high-quality products are delivered (Office of Government Commerce, 2009).

Various authors and institutions define project auditing, which can be summarised as the process of reviewing the management of a project, evaluating project results, and the project's compliance with project management standards in order to ensure project success (Hill, 2007; Huemann, 2004; Stanleigh, 2009). The areas of common understanding in project auditing are:

- ***Examination:*** The assessment or review of the management of a project and evaluation of project progress against the project schedule helps to determine factors contributing to project results.
- ***Project management:*** Project auditing assesses the management of a project, including its methodology as well as project results, against the project schedule.
- ***Compliance:*** Project management is verified to determine if it complies with the various standards, as discussed in Chapter 6.

10.7.2 Types of Project Audits

Various types of project audits are highlighted within the project management literature. The following types of project audits can be performed by the project manager:

- ***Pre-project audit:*** This audit validates the project's readiness to initiate and facilitate the transition from the project planning phase to the project execution phase. The purpose is to determine the viability of achieving the various project activities as well as the implementation strategy (Duffy & Thomas, 1989).
- ***Project health-check audit:*** This audit assesses the performance of a project during the project execution phase. The aim is to review the ongoing project execution, diagnose any problems, and recommend improvements (Hill, 2007).
- ***Post-project audit:*** This type of audit is conducted after the closing phase of the project life cycle to determine whether the project can be formally closed. The purpose of this audit is to define and record lessons learnt to

improve the performance of future projects (Hill, 2007). It also reviews the benefits realisation of the implemented project.

- ***Technical audit:*** This audit aims at evaluating deficiencies or areas of improvement in a process or system. Technical audits are conducted during the planning phase in conjunction with the pre-project audit (Hill, 2007).
- ***Customer satisfaction audit:*** This audit involves the customer and focuses on the customer's perception of the project's progress in achieving the desired objectives (Hill, 2007). This audit should be performed during the health-check and post-project audits.
- ***Project recovery audit:*** This audit is similar to the combined content of the project management audit and the project performance audit but with a focus on the indicators of unsatisfactory project performance (Hill, 2007).
- ***Project resource utilisation audit:*** This audit examines the fulfilment of resource allocation and the timely assignment of resources necessary for the accomplishment of specific tasks. It also examines the effectiveness of project managers in assigning resources during the implementation of the project in order to meet project objectives. This audit does not examine the performance of resources (Hill, 2007).
- ***Project team performance audit:*** Auditing the performance of the project team involves reviewing project work assignments and ensuring that they are aligned with the team members' technical and professional competencies. Auditing the behaviour of the project team is an important aspect because the project team is critical to the success of the overall project (Hill, 2007).
- ***Contractor audit:*** This audit forms part of the project procurement management knowledge area and audits contractors within the project environment (Hill, 2007). This audit includes the contractor management plan as well as the contract management review.
- ***Project management methodology audit:*** This audit examines the use of the selected PM methodology and validates its content and effectiveness. It is an audit that transcends individual projects and project managers to gain a perspective of the application on project management processes and practices across all projects within the relevant organisation (Hill, 2007).
- ***Compliance audit:*** This audit provides a comprehensive review of the project's adherence to regulatory guidelines. This type of audit is performed by independent accounting, security, or IT professionals who evaluate the strength and thoroughness of compliance preparations (Hill, 2007).

The project manager, in consultation with the organisational leaders, should determine what type of project audit should be performed. Project audits should not be conducted as a way to punish the project manager and project team but rather as a way to improve future performance. The concern with project audits is that they are often done after the fact, when it is sometimes too late to rectify a problem or situation. Project managers should recognise the value of the various types of audits and set time aside in the project schedule to perform these audits at different times.

10.7.3 Framework for Project Auditing

Project auditing should be a continuous process that actually focuses on assurance rather than on auditing. A framework for continuous project auditing is illustrated in Figure 10-2.

The auditing of a project is guided by project governance and should be done within this mindset. The first level of audits occurs at the project life cycle. The aim is to determine whether the project manager adhered to the processes of the project life cycle. During the project life cycle, various project deliverables are created. The purpose of the audit at level two is to determine whether these various deliverables are adhering to requirements, specifications and quality.

Level three is where project deliverables are audited based on the project life cycle. These audit phases examine the project deliverables from each phase of the project life cycle. The pre-audit phase examines the project deliverables from the feasibility phase of the project life cycle. The basic project deliverables include the business case and the project feasibility study report. The mid-audit phase examines project deliverables from the initiation, planning, execution, monitoring and controlling, and closing phases of the project life cycle. The post-audit examines project deliverables from the benefits sustainment phase of the project life cycle.

The results from these audits are used at the various gate reviews to determine whether the project is still viable and whether the project should continue to the next phase. The gate reviews should also be used to highlight deficiencies that can be rectified during the next phase. The aim should be to encourage the project manager to rectify concerns or issues where possible and not to terminate the project.

10.8 The Executive's Role in Project Governance

Ideally, a positive relationship exists between corporate governance and project success. In order for the governance of projects and project governance to be embedded into project management, executives should ensure that they abide

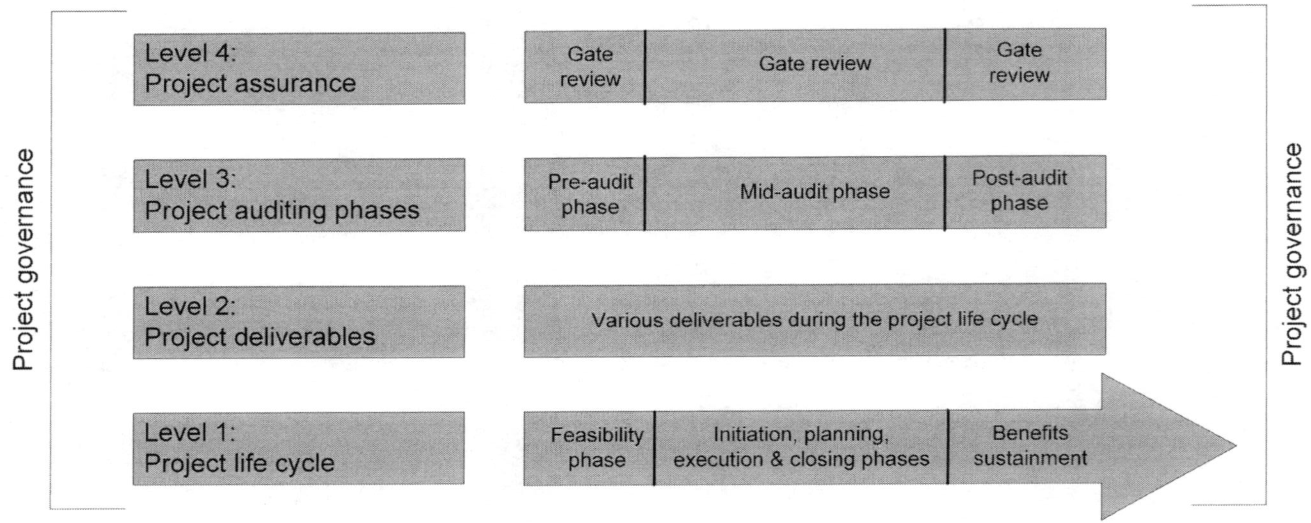

Project governance

Level 4:
Project assurance

Gate review

Gate review

Gate review

Level 3:
Project auditing phases

Pre-audit phase

Mid-audit phase

Post-audit phase

Level 2:
Project deliverables

Various deliverables during the project life cycle

Level 1:
Project life cycle

Feasibility phase

Initiation, planning, execution & closing phases

Benefits sustainment

Project governance

Figure 10-1 Framework for Project Auditing (Adapted from Mkoba and Marnewick [2016])

and adhere to the principles of corporate governance. Project governance should then become a way of working within the organisation. With regard to projects and project management specifically, executives can push the project governance agenda in various ways.

10.8.1 Establishment of Formal Roles and Responsibilities

Project management continues to be a challenging concept to get people to take seriously, as highlighted earlier in the chapter. When the various stakeholders are not even sure what their roles and responsibilities are within project governance, then this becomes an even more confusing topic. Stakeholders might have a slight idea of what their roles and responsibilities are, but it is the duty of the executives to establish and clarify the various project governance roles and the associated responsibilities. This serves two purposes. The first purpose is that it creates clarity, and every stakeholder knows his or her role and responsibilities. The second purpose is that it creates a platform for accountability. Accountability goes hand in hand with the various roles and responsibilities, and executives can enforce corrective action when a stakeholder is neglecting his or her responsibilities.

10.8.2 Walk the Governance Talk

Executives are directly involved in corporate but not project governance. Project governance is a subset of corporate governance, and the behaviour and attitude of executives toward corporate governance directly influences the behaviour and attitude of project managers toward project governance. When the impression is created that executives do not take corporate governance seriously, then project managers will also not take project governance seriously. The adage that behaviour creates behaviour is relevant when it comes to corporate and project governance. When executives walk the corporate governance talk, then project managers will also walk the project governance talk.

10.8.3 Importance of Governmentality

Governmentality is the foundation of project governance. Executives should implement various training and education programmes, highlighting the importance of governmentality as a potential success factor (Müller et al., 2017). The focus of these training and education programmes should be to educate the various project stakeholders (i) on the project governance roles and responsibilities; and (ii) on the importance of walking the project governance talk. The establishment of training and education programmes increases governmentality. This increases the importance of project governance as a factor that contributes to project success.

10.8.4 Training

The notion of governmentality is noble, but it will not be effective when the project manager does not understand the various standards, legislation, and reports. The purpose of the training should be to make project managers aware of corporate governance, in general, as well as the environment (organisation or country) in which the project is implemented, in particular. The training should also focus on project governance and how the principles of corporate governance should be incorporated within the project itself. Knowledge is power, and project managers can be empowered by executives if there is a strong emphasis on the training of project governance.

10.8.5 Enforce Continuous Auditing

The auditing of projects strengthens governmentality, which enhances project governance. Executives should address two concerns with regard to project auditing. The first concern is that project auditing is not a nice-to-have but that it is compulsory. The type and depth of the project audit are dependent on the project itself and the strategic importance of the project. The more important a project is, the more emphasis should be placed on auditing the project in all its facets. The second concern is that project auditing is actually a continuous process and should be done throughout the project life cycle. The emphasis is more on project assurance rather than project auditing. This implies that when certain decisions and/or recommendations are done at the gate reviews, as indicated in Figure 10-2, then the executives should abide by these decisions, even when it means that a project must be cancelled.

10.9 Conclusion

This chapter focused on two important aspects, namely, project governance and project auditing. The notion of project governance was unpacked, and the difference between governance of projects, project governance, and governmentality was illustrated. The most important roles and responsibilities were discussed in relation to the project governance framework. Project auditing focused on the importance of continuous auditing or assurance. Various types of project audits were highlighted, indicating that the project manager in consultation with the stakeholders should decide which type of audit is relevant.

Organisational leaders should realise that projects are not implemented in isolation and that they are part and parcel of the larger organisation. Just as corporate governance is important to steer and direct the organisation, project governance is important to steer and direct the project. Project governance provides the framework and rules that a project must abide by, and project auditing determines whether project managers and executives played by the rules. This is the important concept highlighted in this chapter.

Executives should realise that project governance and auditing are important to the overall success of a project. Organisational strategies run the risk of not being implemented properly when too many projects are failing. When project failure is ascribed to the lack of project governance and auditing, then executives should be accountable because they are the ones who should enforce governance, in general.

10.10 References

Badewi, A. (2016). The Impact of Project Management (PM) and Benefits Management (BM) Practices on Project Success: Towards Developing a Project Benefits Governance Framework. *International Journal of Project Management, 34*(4), 761–778.

Bekker, M. C. (2014). Project Governance: "Schools of Thought." *South African Journal of Economic and Management Sciences, 17*(1), 22–32.

Bekker, M. C. (2015). Project Governance—The Definition and Leadership Dilemma. *Procedia—Social and Behavioral Sciences, 194*, 33–43.

Bekker, M. C., & Steyn, H. (2008, 27–31 July). *The Impact of Project Governance Principles on Project Performance.* Paper presented at the PICMET '08—2008 Portland International Conference on Management of Engineering & Technology, Cape Town, South Africa.

Brunet, M., & Aubry, M. (2016). The Three Dimensions of a Governance Framework for Major Public Projects. *International Journal of Project Management, 34*(8), 1596–1607.

Cadbury, A. (1992). *Cadbury Report: The Financial Aspects of Corporate Governance.* Retrieved from http://www.ecgi.org/codes/documents/cadbury.pdf

Cadbury, A. (2002). *Corporate Governance and Chairmanship: A Personal View.* London, UK: Oxford University Press.

Clarke, T., & Branson, D. (2012). *The SAGE Handbook of Corporate Governance.* London, UK: SAGE Publications Ltd.

Duffy, P. J., & Thomas, R. D. (1989). Project Performance Auditing. *International Journal of Project Management, 7*(2), 101–104.

Hill, G. M. (2007). *The Complete Project Management Office Handbook* (2 ed.). Boca Raton, FL, USA: Auerbach Publications.

Huemann, M. (2004). *Management Audits of Projects and Programmes—How to Improve Project Management and Programme Management Quality.* Paper presented at the III IPMA ICEC International Expert Seminar, Bilboa, Spain.

Institute of Directors Southern Africa. (2016). *King IV Report on Corporate Governance.* Retrieved from https://c.ymcdn.com/sites/iodsa.site-ym.com/resource/collection/684B68A7-B768-465C-8214-E3A007F15A5A/IoDSA_King_IV_Report_-_WebVersion.pdf

Iskander, M. R., & Chamlou, N. (2000). *Corporate Governance: A Framework for Implementation* (20829). Retrieved from http://documents.worldbank.org/curated/en/810311468739547854/pdf/multi-page.pdf

Joslin, R., & Müller, R. (2016). The Relationship between Project Governance and Project Success. *International Journal of Project Management, 34*(4), 613–626.

Lemke, T. (2015). Foucault, Governmentality, and Critique. *Rethinking Marxism, 14*(3), 49–63.

Maurizio, A., Girolami, L., & Jones, P. (2007). EIA and SOA: Factors and Methods Influencing the Integration of Multiple ERP Systems (in an SAP environment) to Comply with the Sarbanes–Oxley Act. *Journal of Enterprise Information Management, 20*(1), 14–31.

McDonald, J. (2002). Software Project Management Audits––Update and Experience Report. *Journal of Systems and Software, 64*(3), 247–255.

Mkoba, E., & Marnewick, C. (2016). *IT Project Success: A Conceptual Framework for IT Project Auditing Assurance.* Paper presented at the Proceedings of the Annual Conference of the South African Institute of Computer Scientists and Information Technologists, Johannesburg, South Africa.

Moeller, R. R. (2008). *Sarbanes–Oxley Internal Controls: Effective Auditing with AS5, CobiT and ITIL.* Hoboken, NJ, USA: John Wiley & Sons, Inc.

Müller, R. (2009). *Project Governance.* Surrey, England: Gower Publishing Limited.

Müller, R., Zhai, L., & Wang, A. (2017). Governance and Governmentality in Projects: Profiles and Relationships with Success. *International Journal of Project Management, 35*(3), 378–392.

Müller, R., Zhai, L., Wang, A., & Shao, J. (2016). A Framework for Governance of Projects: Governmentality, Governance Structure and Projectification. *International Journal of Project Management, 34*(6), 957–969.

OECD. (2015). *G20/OECD Principles of Corporate Governance.* Paris, France: OECD Publishing.

Office of Government Commerce. (2009). *Managing Successful Projects with PRINCE2* (5 ed.). United Kingdom: The Stationary Office.

Project Management Institute. (2013). *A Guide to the Project Management Body of Knowledge (PMBOK® Guide)* (5 ed.). Newtown Square, PA, USA: Project Management Institute.

Project Management Institute. (2016). *Governance of Portfolios, Programs, and Projects: A Practice Guide.* Newtown, PA, USA: Project Management Institute.

Reusch, P. (2011, 15–17 September). *New Standards for Project Audit and the Impact on Existing Standards for Project Management.* Paper presented at the Proceedings of the 6th IEEE International Conference on Intelligent Data Acquisition and Advanced Computing Systems, Prague, Czech Republic.

Stanleigh, M. (2009). *Undertaking a Successful Project Audit.* Retrieved from https://www.projectsmart.co.uk/undertaking-a-successful-project-audit.php

Too, E. G., & Weaver, P. (2014). The Management of Project Management: A Conceptual Framework for Project Governance. *International Journal of Project Management, 32*(8), 1382–1394.

Turner, J. R. (2006). Towards a Theory of Project Management: The Nature of the Project Governance and Project Management. *International Journal of Project Management, 24*(2), 93–95.

Chapter 11

Executive Sponsor

~ Ask not what your country can do for you, ask what you can do for your country. ~

— John F. Kennedy*

The previous chapters focus exclusively on the discipline of project management. The purpose was to position project management within the organisation and demonstrate the value that project management brings to the organisation. Each chapter concludes with a section on what executives can do in an indirect way to assist project managers in ensuring that projects are a success. The focus of this chapter is on organisational leaders and, specifically, the role they play as executive sponsors for projects. The executive sponsor is part and parcel of the project and contributes directly to the success of the project, rather than merely a bystander offering advice. It was found that an actively engaged executive sponsor is a top driver for project success (Project Management Institute, 2014b).

The first section of this chapter defines the executive sponsor and distinguishes between the concepts of executive sponsor and project champion. The second section focuses on the characteristics that an executive sponsor must display. This is followed, in the third section, by a discussion on the governance role as well as the control and support functions that the executive sponsor fulfils. The fourth section determines the positive relationship between project success and an involved executive sponsor. Section five briefly investigates the challenges that an executive sponsor faces. And the sixth section focuses on how executive sponsors can be empowered.

* Retrieved from https://www.brainyquote.com/quotes/quotes/j/johnfkenn109213.html

11.1 Defining the Executive Sponsor

The executive sponsor provides resources and support for a specific project and is accountable for the ultimate success of the project (PMI, 2013). The Association for Project Management (2006) shares this view but emphasises that the executive sponsor is also responsible for the identification of the business need or problem. If this view relates back to Chapter 6 in which project management was discussed, then it implies that the executive sponsor is involved in the project from the conceptualisation phase right through to benefits realisation and sustainment. Lechtman (2005) is also of the opinion that the executive sponsor should have political and financial power as well.

A distinction must be made between the executive sponsor and the project champion. These two roles are distinct from each other and should not be used interchangeably. The executive sponsor can be seen as the owner of the project, whereas the project champion is perceived as someone who advocates the benefits of the project (Crawford & Brett, 2001). Any stakeholder in the project should thus be a champion of the project, but only one individual can be the executive sponsor. The executive sponsor can, however, be a champion as well. Cooke-Davies (2005) is of the opinion that being an executive sponsor

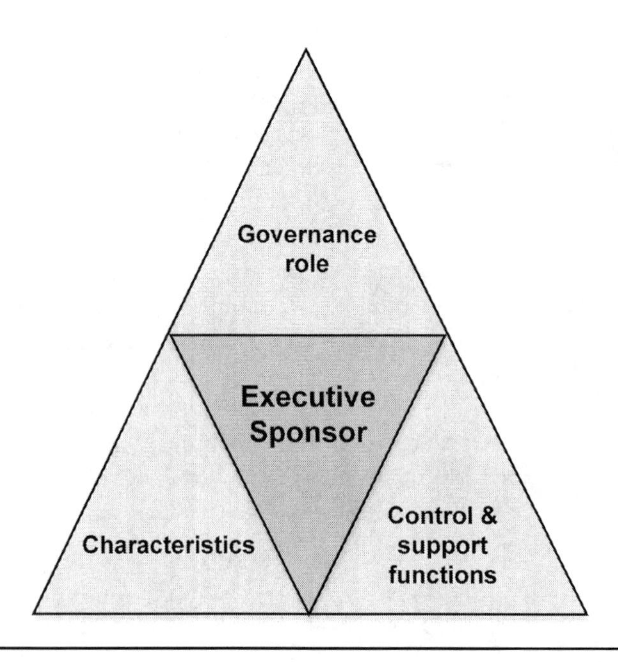

Figure 11-1 Executive Sponsor

incorporates the role of the project champion. The role of the executive sponsor is formalised, whereas the role of the champion is more of an informal nature.

The composition of an executive sponsor is fairly complex, and three major components or aspects of the executive sponsor can be derived from extant literature. These three components, as illustrated in Figure 11-1, comprise (i) the governance role that the executive sponsor fulfils; (ii) the control and support function the executive sponsor fulfils; and (iii) the characteristics that the executive sponsor must own.

The first component or aspect that needs to be unpacked is the executive sponsor's characteristics. The executive sponsor should demonstrate certain characteristics to be successful in this role as well as to ensure project success.

11.2 Characteristics of the Executive Sponsor

Various characteristics are assigned to the executive sponsor, the most important of which are discussed in this section (Crawford et al., 2009; Kloppenborg, Manolis, & Tesch, 2009). Figure 11-2 provides a synopsis of the characteristics an executive sponsor should display.

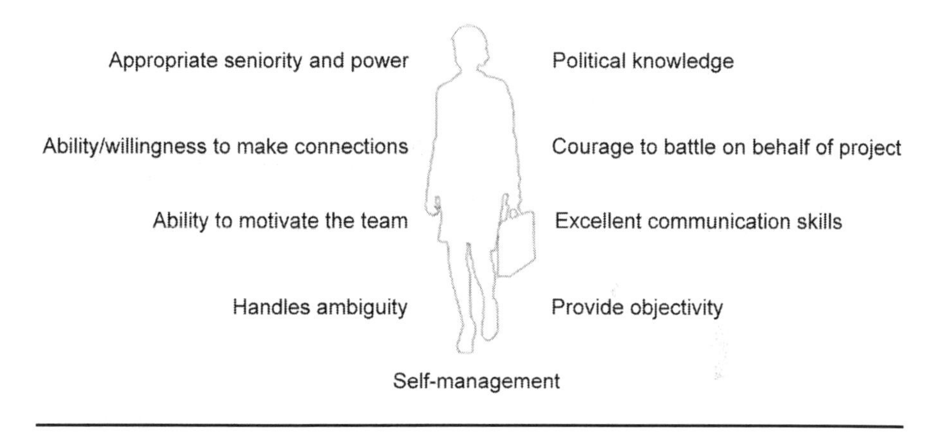

Figure 11-2 Characteristics of the Executive Sponsor

- ***Appropriate seniority and power:*** The title of executive sponsor implies a certain level of seniority and associated power. Seniority provides the executive sponsor with certain knowledge and wisdom about the organisation itself, peers and the environment within which the organisation is functioning. Having the appropriate seniority without the associated power is not worth much to the executive sponsor. Seniority must be associated with the capability to direct or influence the behaviour of other senior executives and the project's stakeholders. This power or influence should be used to persuade others of the benefits of the project and to remove certain obstacles that might have a negative impact on the success of a project.

- ***Political knowledge:*** Schwalbe (2016) describes political knowledge as the knowledge that the executive sponsor should have about competition among divisions, groups, and individuals. This competition is centered around power, resources, and leadership. Political knowledge empowers the executive sponsor, who predetermines the actions and responses of individuals based on their own quest for power, resources, and leadership.
- ***Ability or willingness to make connections:*** One of the most important characteristics that the executive must display is the ability to network. Without a network that can support and influence decisions, the executive sponsor will have a difficult time pushing the project's agenda. This ability or willingness is not necessarily associated with seniority and power. The executive sponsor should determine whether this is a personal weakness and undergo the necessary mentoring to strengthen this characteristic.
- ***Courage to battle on behalf of the project:*** An organisation can sometimes be perceived as a war zone, in which each and everyone is fighting for his or her own interest. In this war zone, the executive sponsor must be prepared to fight for the good of the project. This battle on behalf of the project should not be just another battle, but it should be based on what is best for the organisation at large. Emphasis should be on the realisation of the organisation's vision and strategies, benefits realisation, and adding value. All of this should be happening within the culture of corporate and project governance.
- ***Ability to motivate the team:*** This ability focuses on the short as well as the long term. The focus of motivating the team in the short term is to encourage them to continue doing their job through issues and setbacks. In the long term, the focus is more on keeping them focused on the project's objectives and the role they play in fulfilling the organisational strategies. This is especially important when a project stretches over a couple of years.
- ***Excellent communication skills:*** Communication skills are the most sought after qualities that organisational leaders look for in individuals. This is even more applicable to the executive sponsor, for whom both vertical and horizontal communication are essential. Reading, writing, and listening carefully are the three most important communication skills. Communication skills can be perceived as the foundation for the other characteristics.

- ***Provide objectivity:*** The executive sponsor should always be level-headed. The project manager and team members are involved in the management of the project and do not always see the bigger picture when it comes to certain decisions or options. The executive sponsor should provide objectivity on the basis of what is best—first for the organisation and second for the project.
- ***Handle ambiguity:*** Projects by nature include ambiguity and uncertainty. The executive sponsor should provide direction and make decisions in the face of ambiguity. The project might experience challenges when the executive sponsor does not have the ability to handle ambiguity. Not everything is always crystal clear in a project, and this characteristic provides the executive sponsor with the ability to draw on previous experience and knowledge in order to address ambiguity.
- ***Self-management:*** The executive sponsor has a part-time role. The executive sponsor is already performing another full-time official role. The ability to self-manage time between a full-time role and the part-time role of executive sponsor is important. Time management plays a major role in self-management. Crawford et al. (2009) believe that the executive sponsor must realise the importance of the executive sponsor role, and only then can efficient self-management take place.

The characteristics of the executive sponsor aid him or her in performing this role in an efficient and effective manner.

11.3 Role of the Executive Sponsor

The executive sponsor has various roles and responsibilities to fulfil. Extant literature describes a wide variety of roles and responsibilities, which are divided into two groups—governance and control (Cooke-Davies, 2005; Crawford et al., 2009)—with an external or internal focus (Bryde, 2008; Crawford & Brett, 2001). Given all the various roles that the executive sponsor must fulfil, it is imperative that these roles are clearly defined and that there is no confusion as to who does what between the executive sponsor and the project manager (Bryde, 2008).

11.3.1 Governance Roles

The governance role focuses on providing guidance on doing the right thing, as per Chapter 10. Specific governance roles that the executive sponsor fulfils include the following:

- ***Serves as the business case owner:*** A business case is a formal document that sets out the rationale and justification for a project investment with the goal of obtaining management commitment and authorisation to proceed (Marnewick, 2014; Ward & Daniel, 2012). The business case is subject to review and ongoing viability testing, throughout the project's lifetime. The executive sponsor is the owner of the business case and is therefore the person who provides the rationale and commitment for any investments.
- ***Provides project objectives:*** The executive sponsor, as the owner of the business case, also provides the project objectives. These project objectives are aligned to the various organisational strategies. The executive sponsor is generally accountable for the realisation of the organisational strategies and is in the best position (i) to provide the project objectives; and (ii) to ensure that the project objectives are achieved.
- ***Harvests benefits:*** The business case promises benefits to the organisation at large and, as such, the executive sponsor is the direct beneficiary of any benefits that are realised and sustained. It is therefore important that the executive sponsor is involved in the benefits realisation management (BRM) process, as described in Chapter 8.
- ***Serves as a key governor of the project:*** Chapter 10 explains the importance of project governance and the various roles people play in it. The executive sponsor is one of the important role players who has a positive impact on project governance. Please refer to Section 10.4.2 for additional information.
- ***Is a "friend in high places" to the project manager:*** One of the characteristics is that the executive sponsor should be able to make connections. According to Cooke-Davies (2005, p. 3), the executive sponsor is "ideally situated to assist with the management of high-ranking stakeholders." This implies that the executive sponsor should use his or her clout and influence to the benefit of the project and the realisation of the benefits.
- ***Supports the project politically:*** Projects require various resources and are sometimes met with resistance from other project managers, executive sponsors, and organisational leaders. This resistance can only be dealt with by the project's own executive sponsor, which implies that the project must be politically supported as well. This political support implies that the executive sponsor must defend the project against those of other organisational leaders and senior executives and convince them of the importance of the project's deliverable to the organisation at large.
- ***Responsible for project's scope:*** As the owner of the business case, the executive sponsor knows what needs to be delivered by the project team to ensure the realisation of the benefits. Therefore, the executive sponsor is directly involved in determining the project's scope and any deviation

from the scope must be authorised by the executive sponsor because a deviation implies fewer benefits.

Apart from the governance roles, the executive sponsor also has control and support functions.

11.3.2 Control and Support Functions

The focus of the governance role is to ensure that the right things are done. The control and support functions focus on doing things right. The executive sponsor fulfils the following functions within these functions:

- ***Provides leadership:*** Leadership means many things to many stakeholders, but one aspect of leadership is that the executive sponsor should direct the project manager and team in the right direction—that is, linked to the strategies and project objectives. And when the project manager deviates from this course, the executive sponsor should provide leadership and guide the team back in the right direction.
- ***Is available:*** Being available is a supportive rather than a control function. The executive sponsor should be available to the project manager. Being available implies that the executive sponsor is committed to the project, and it creates a positive feeling among the project team and stakeholders if the executive sponsor is available during scheduled as well as unscheduled times. Availability also ensures the quick resolution of issues and timely decision making.
- ***Provides budget allocation:*** The budget for the project is the responsibility of the executive sponsor. The resources needed to implement the project are in the business case. A specific resource is money, and a budget will be allocated to the project. The project manager is responsible to manage the project within the budget (PMI, 2013), but the executive sponsor is responsible for sourcing the money and allocating the money to the project.
- ***Makes major decisions for the project:*** Project managers do not necessarily have the bigger picture of how a project contributes to the organisational strategies. Major decisions that impact the realisation of benefits and the implementation of organisational strategies should be made by the executive sponsor. The rationale is that the executive sponsor is ultimately the owner of the project deliverables as well as the realised benefits. It therefore makes sense that the executive sponsor is then accountable for making major decisions.
- ***Finds resources for the project:*** Resources are a scarce commodity, especially when one is referring to human resources. Project Resource

Management is a knowledge area in the *PMBOK® Guide* (PMI, 2017), and the project manager is responsible for the management of resources. The problem arises when there are no resources to manage. This is where the executive sponsor plays an important role in finding the necessary skilled resources, internal or external to the organisation. The executive sponsor makes use of his or her seniority and various networks to secure the necessary resources.

- ***Approves the project schedule:*** Benefits realisation and the implementation of organisational strategies are directly linked to the time that the project deliverable is delivered. The project manager creates the project schedule, but the executive sponsor finally approves the schedule based on when benefits should be realised. The approval should be based on a realistic and achievable project schedule, and the executive sponsor should not use his or her power to bully the project manager into unrealistic time frames.
- ***Ratifies decisions made by the project manager or team:*** This function ties in with the availability function wherein the project manager and team use the executive sponsor to ratify decisions. The ratification of decisions is not possible if the executive sponsor is not available. The purpose of this function is to ratify decisions, with the success of the project as the goal.

It is evident that the role of the executive sponsor is complicated and requires daily involvement. The executive sponsor must have a hands-on attitude and cannot keep a distance. The continuous involvement of the executive sponsor contributes to project success, as indicated in the next section.

11.4 Project Success and the Executive Sponsor

A positive relationship exists between the success of a project and the level of involvement of the executive sponsor (Kloppenborg et al., 2009). Various studies highlight that the support of top management contributes to successful projects (Bryde, 2008; Marnewick, 2013). Few studies actually emphasise what the executive sponsor must do to improve the success of a project. Apart from performing the roles and functions as stipulated earlier, Labuschagne and Lechtman (2006) identify control objectives that the executive sponsor should adhere to for project success. Many of these control objectives are linked with the roles and functions, as described below.

These control objectives can be summarised as follows: First, the executive sponsor should understand what it ***implies*** to be an executive sponsor, what the role entails, and perform the role to the best of his or her abilities. Second, the executive sponsor should ***set time aside*** to be a dedicated sponsor. This ties in with the role of being available to the project manager and team. The executive sponsor will be extremely involved during the initiation and benefits realisation phases but less so during the execution phase. The executive sponsor must also be up to date with the ***project status***. This implies continuous communication between the executive sponsor and the project manager. One of the roles of the executive sponsor is to make major decisions and determine the direction of the project. Determining and setting the ***project direction*** is a contributor to project success. Project managers often require resources and skills beyond those of the project team that they are managing. Should the project manager not be able to ***access resources and skills*** from within the organisation, then the executive sponsor may be called upon to assist the manager in obtaining them. The ***project must be closed*** properly before the executive sponsor can take ownership of the deliverables and the associated benefits. The closure should also be used to determine the success of the project. The project success criteria discussed in Chapter 6 can be used to determine the success of the project at closeout. The executive sponsor should continuously ***appraise the project***. This should happen throughout the project life cycle. The purpose of this appraisal is to determine the viability of the project and if there is still value in executing the project. This links with project auditing and assurance as discussed in Section 10.7.3. The executive sponsor is ***accountable for the business case***, and should the project fail, the sponsor would be held personally liable for any losses incurred. The same rationale applies where the executive sponsor is then also ***accountable for the realisation of the benefits***. The business case and benefits realisation go hand in hand, and the executive sponsor promised that certain benefits will be realised if the project is a success. The executive sponsor is seen as the chairperson of the steering committee and, as such, has the power to appoint the steering committee members. The executive sponsor ***represents the project*** on the steering committee. The executive sponsor should look after the ***interests of all the stakeholders***. He or she must also ensure that all stakeholder interests are aligned with the project objectives. The executive sponsor is particularly interested in business risks and issues that may arise should the project fail to deliver according to the business case. As such, the executive sponsor should ensure that the project has appropriate ***risk management*** processes in place.

The role of the executive sponsor is not always plain sailing. Certain challenges face the executive sponsor, which can have a negative impact on project success.

11.5 Challenges Facing the Executive Sponsor

Three major challenges face the executive sponsor on a day-to-day basis: over-extension, inefficient communication, and a lack of professional development (PMI, 2014b).

1. ***Overextension:*** The first challenge that an executive sponsor faces is overextension. This implies that apart from performing his or her day-to-day roles, the executive sponsor must also perform executive sponsor roles. When the executive sponsor becomes too involved in too many projects, then he or she becomes overextended. The result is that sufficient attention cannot be given to either the day-to-day functional role or the executive sponsor role. Research done by PMI highlights that executive sponsors are overextended most of the time (PMI, 2014b).
2. ***Inefficient communication:*** It is important for an executive sponsor to have excellent communication skills. However, an executive sponsor might display excellent communication skills, but if these skills are applied ineffectively in the project environment, then there is the possibility that he or she cannot adequately fulfil this role. The role of executive sponsor is best performed when he or she communicates effectively with all of the stakeholders. In addition, it is more important for the executive sponsor to communicate effectively with the project manager and other senior executives than with the project team.
3. ***Lack of professional development:*** Research highlights that executive sponsors do not necessarily understand the role that they need to perform and that they are performing this role by accident. Executive sponsors must have knowledge of portfolio, programme, and project management (Crawford & Brett, 2001). Executive sponsors who have this knowledge tend to have better project outcomes (PMI, 2014a). Apart from having P3 knowledge, additional skills and training are needed, as well as experience. This can be obtained through external and internal training as well as through mentoring from experienced executive sponsors. It is imperative that executive sponsors undergo training in the role itself.

Overcoming these three challenges will make executive sponsors more effective and efficient, which will result in more projects with positive outcomes. The following section focuses on the role of organisational leaders in empowering executive sponsors.

11.6 Empowering the Executive Sponsor

By default, in most instances organisational leaders will assume the role of executive sponsor for most of the projects that form part of the larger project portfolio. The challenges described previously highlight that everything is not plain sailing and that a concerted effort is needed to ensure the successful fulfilment of the role of executive sponsor.

11.6.1 Recognise and Define the Executive Sponsor Role

The role of the executive sponsor is not an accidental one. The mere fact that organisational leaders implement projects implies that there will be executive sponsors. The first step in empowering executive sponsors is to recognise the role they play and the importance of this role in realising organisation strategies. This role is not a formal role that slots in the organisational hierarchy, but it is one of those roles that needs to be recognised. It might be best suited that this role forms part of the project management office (PMO). The PMO should also define the role of the executive sponsor based on the governance and control aspects of the role. Defining the role of the executive sponsor immediately formalises and empowers it. This, in turn, provides structure and accountability to this role. At the end of the day, executive sponsors should be accountable for the projects that they sponsor; they will either cheer the successes or weep for the losses.

11.6.2 Provide Guidance and Training for Executive Sponsors

One of the challenges that face an executive sponsor is a lack of project management knowledge (PMI, 2014b). Guidance and training to address this knowledge gap hinges on two aspects. The first aspect of the training should be on portfolio, programme, and project management. The focus should be generalised in order for the executive sponsor to understand what these roles entail and how portfolios, programme, and projects are generally managed. The second aspect of the training should address the role of executive sponsor and what is expected from its incumbent. The training should focus on the characteristics as well as the governance and control dimensions of this role.

11.6.3 Acceptance of the Role

Understanding the role of the executive sponsor is not enough. Executive sponsors should also accept this role. By accepting this role, the executive sponsor acknowledges that he or she accepts the responsibilities as well as the accountability associated with it. The acceptance of the role of executive sponsor should be formalised and facilitated by the PMO. Acceptance of the role is implied when a person sponsors a project, but this is not sufficient. An implied acceptance is not enough because it will be easier for the executive sponsor to neglect the responsibilities inherent in this role if it is not formalised. Therefore, no formal accountability is associated with this implied acceptance. The role must therefore be officially accepted because this creates a formal structure.

11.6.4 Change of Project Sponsor

It might happen that the executive sponsor of a project is changed. This change can be attributed to various reasons, including the resignation of the executive sponsor. Whatever the reason, the change of an executive sponsor creates challenges for the project. The new executive sponsor might not have the same dedication to the project because he or she was not the one championing the project from the start and might not feel the same about the project as the original incumbent. The new executive sponsor must really make an effort to understand the rationale behind the implementation of the project, and the contribution and benefits the project's deliverable will have for the organisation at large. Organisational leaders should anticipate that executive sponsors might change and have contingency plans in place to consider and address potential difficulties (Crawford & Brett, 2001).

11.7 Conclusion

This chapter investigated the roles that organisational leaders play as executive sponsors for projects. The role of executive sponsor was defined with a focus on the characteristics of an executive sponsor as well as the governance roles and control functions that an executive sponsor must perform on a daily basis. Various roles and functions were attributed to the executive sponsor, and it is evident that fulfilling this duty can be quite daunting and challenging. A positive relationship between project success and executive sponsor involvement was established, with control objectives that can be used to determine whether the executive sponsor is diligently performing the duties inherent in this role.

The executive sponsor plays an important role in the portfolio, programme, and project management environment. This is not a position that needs to be filled quickly. On the contrary, effort and thought should be given as to who should fulfil this duty. This person's contribution to project success and the ultimate realisation of the benefits should not be underestimated. This person plays a vital role in the bigger picture, and without an effective and efficient executive sponsor, the project is going to face serious challenges. It is also evident that the executive sponsor must be serious about this position and that serious thought should be given before taking on this responsibility.

The focus of the executive sponsor should be on what can be done for the project at large, thus ensuring that the project is successful. Focusing on what is best for the project will result in benefits for the executive sponsor, the project manager, the project team, and the organisation at large.

11.8 References

Association for Project Management. (2006). *APM Body of Knowledge* (5 ed.). Buckinghamshire, UK: Association for Project Management.

Bryde, D. (2008). Perceptions of the Impact of Project Sponsorship Practices on Project Success. *International Journal of Project Management, 26*(8), 800–809.

Cooke-Davies, T. J. (2005). *The Executive Sponsor—The Hinge Upon Which Organisational Project Management Maturity Turns.* Paper presented at the PMI Global Congress, Edinburgh, Scotland.

Crawford, L., Cooke-Davies, T. J., Hobbs, B., Labuschagne, L., Remington, K., & Chen, P. (2009). *Situational Sponsorship of Projects and Programs: An Empirical Review.* Newtown Square, PA, USA: Project Management Institute.

Crawford, L. H., & Brett, C. (2001). *Exploring the Role of the Project Sponsor.* Retrieved from http://citeseerx.ist.psu.edu/viewdoc/download?doi=10.1.1.555.6028&rep=rep1&type=pdf

Kloppenborg, T. J., Manolis, C., & Tesch, D. (2009). Successful Project Sponsor Behaviors during Project Initiation: An Empirical Investigation. *Journal of Managerial Issues, 21*(1), 140–159.

Labuschagne, L., & Lechtman, E. (2006). *Sponsoring Projects from a Project Governance Perspective.* Paper presented at the Proceedings of the 2006 PMSA International Conference, Johannesburg, South Africa.

Lechtman, E. (2005). *A Holistic Framework for Successfully Sponsoring IT Projects from an IT Governance Perspective.* (M.Com. Dissertation), Rand Afrikaans University, Johannesburg, South Africa. Retrieved from http://hdl.handle.net/10210/282

Marnewick, C. (2014). The Business Case: The Missing Link between Information Technology Benefits and Organisational Strategies. *Acta Commercii, 14*(1), 1–11.

Marnewick, C. (Ed.) (2013). *Prosperus Report—The African Edition.* Johannesburg, ZA: Project Management South Africa.

Project Management Institute. (2013). *A Guide to the Project Management Body of Knowledge (PMBOK® Guide)* (5 ed.). Newtown Square, PA, USA: Project Management Institute.

Project Management Institute. (2014a, November). Engaging Sponsors. *PM Network, 28,* 18–20.

Project Management Institute. (2014b). *Executive Sponsor Engagement—Top Driver of Project and Program Success.* Retrieved from http://www.pmi.org/-/media/pmi/documents/public/pdf/learning/thought-leadership/pulse/executive-sponsor-engagement.pdf

Project Management Institute. (2017). *A Guide to the Project Management Body of Knowledge (PMBOK® Guide)* (6 ed.). Newtown Square, PA, USA: Project Management Institute.

Schwalbe, K. (2016). *Information Technology Project Management* (8 ed.). Boston, MA, USA: Cengage Learning.

Ward, J., & Daniel, E. (2012). *Benefits Management: How to Increase the Business Value of Your IT Projects* (2 ed.). Chichester, UK: John Wiley & Sons.

Chapter 12

Comprehensive Overview of Project Management

~ What we call the beginning is often the end. And to make an end is to make a beginning. The end is where we start from. ~

— T.S. Elliot*

The purpose of this chapter is not to provide a summary of the previous chapters, but to provide a comprehensive overview of what constitutes project management and how organisational leaders can derive value from it. In the creation of a comprehensive overview, one realises that project management itself is complex and that various concepts play a positive role in the creation of value. The rationale for a comprehensive overview is that the various concepts discussed in the previous chapters are discussed in isolation, with limited reference to the other concepts. Once all the various concepts are collated into one comprehensive overview, one starts to appreciate the complexity surrounding project management, and the impact project management has on the organisation at large.

The various concepts discussed in the earlier chapters do not necessarily fit easily together like LEGO® blocks. This comprehensive overview provides a best-fit scenario and fits these concepts together logically. This logical fit might not be applicable to all organisations all of the time but it provides a rational connection between various concepts. It is also possible that some concepts are

* Retrieved from https://www.brainyquote.com/quotes/quotes/t/tseliot101421.html

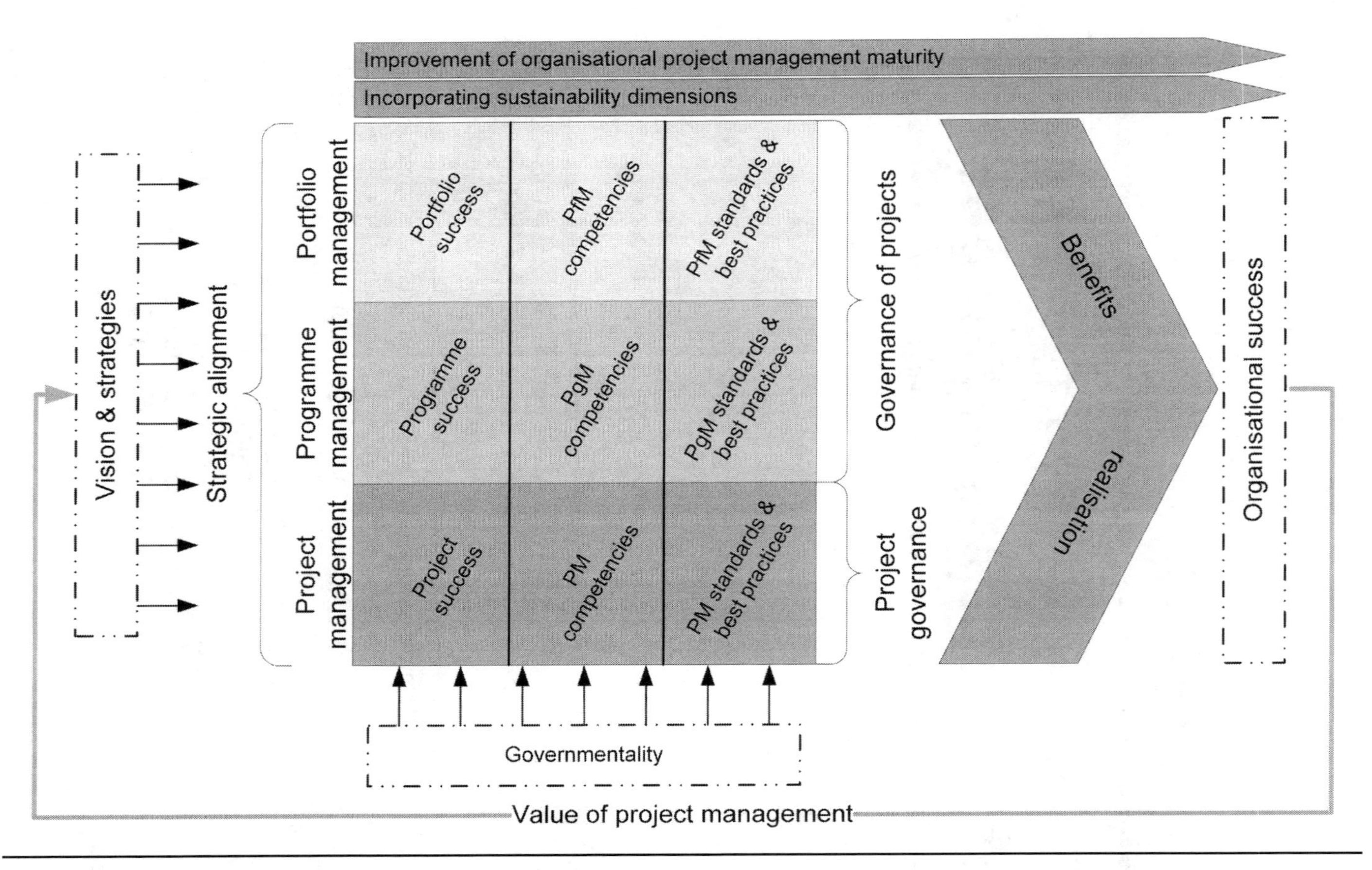

Figure 12-1 Comprehensive 3PM Framework

not applicable to all organisations. Some smaller organisations might not see the purpose of having programmes and programme management. In these cases, they can then just discard these concepts from the comprehensive overview.

12.1 Comprehensive 3PM Framework

A graphical representation of the comprehensive framework is presented in Figure 12-1.

It is evident that the concept of project management cannot be dealt with in isolation. Project management—that is, the management of projects—is influenced by other concepts and, in turn, influences these concepts as well. There is a symbiotic relationship between all the different concepts, and when something is going wrong within one concept, it might negatively influence other concepts.

12.2 Symbiotic Relationships

The concept that gives life to every other concept in this framework comprises the vision and strategies of the organisation. The vision and strategies are acting as the rudder of a boat and provide the direction of the portfolio, programmes, and projects. Executives and organisational leaders should ensure that the vision and strategies are continuously communicated, as discussed in Section 3.8.1. Portfolios, programmes, and projects should be strategically aligned to the organisational vision and strategies. The V2P framework can aid organisational leaders to derive possible programmes and projects from the vision and strategies.

As the rudder of a ship provides direction, the sails provide the additional means to manoeuvre. Manoeuvring implies that the ship can go in various directions at different speeds. The same applies to 3PM. Portfolio, programmes, and projects are perceived as the means to implement the strategies and ultimately the vision. Portfolio management provides the optimum way to implement programmes and projects. Programmes and projects can be changed as the needs of the portfolio change. This is all dependent on the vision and strategies. When the strategies are changed, then the portfolio needs to be adjusted, which has an impact on the programmes and projects within the portfolio.

Organisational leaders must be skilled and aware of the various generic concepts that form part of portfolio, programme, and project management. These concepts include but are not limited to (i) success; (ii) competencies; and (iii) standards.

- ***Success:*** The success of a portfolio, programme, and project is determined and measured differently. Organisational leaders must be aware of the following: (i) the role that they themselves play to ensure success at all three levels; (ii) how to determine success up front; and (iii) how to accurately measure success in an unbiased fashion. This implies that organisational leaders should be able to determine success at a strategic as well as operational level.
- ***Competencies:*** Various competencies are needed to manage a portfolio, programme, or project. A competent project manager does not imply a competent programme manager. Various skill sets are needed for these three levels, and organisational leaders must ensure that the correct education and training are provided to portfolio, programme, and project managers. Having the wrong person with the wrong competencies can be disastrous for the organisation.
- ***Standards and best practices:*** The third concept focuses on standards and best practices. Standards and best practices provide the foundation for how portfolios, programmes, and projects should be managed. Using the analogy of a ship again, there are certain ways that knots are tied and instruments are used. This is universal to all ships, and everyone manning a ship recognises it. The same applies to 3P management. When standards and best practices are in place, then everyone realises what needs to be done and by whom it must be done.

An underlying principle that needs to be factored in during portfolio, programme, and project management is the notion of governmentality. Chapter 10 discussed the notion of governmentality in detail. When the correct perceptions, attitudes, values, and culture are in place, then doing the right thing becomes second nature. People involved in portfolios, programmes, and projects will, by default, know and understand what needs to be done, and what the correct way is to perform their duties and obligations.

Governmentality itself is unfortunately not the answer, and governance structures need to be in place. The first concept is the governance of projects. This plays a role at the portfolio and programme levels, and the focus is to ensure that the vision and strategies are implemented through an optimum portfolio. Project governance ensures that the projects are implemented within the guidelines and structures created by the project management office (PMO). The comprehensive 3PM framework cannot function properly without incorporating the notion of governance. Organisational leaders should promote governance through active engagement, as stipulated in Section 10.8.

The successful management of a portfolio, programmes, and projects culminates in the realisation of benefits, which is the ultimate goal of implementing projects. Benefits are identified during the inception of a project. Benefits act as the compass, and any deviations should be addressed through the adjustment of the sails—that is, the portfolio, programmes, and projects. Managing benefits and ultimately delivering benefits to the organisation ensure the success of the organisation.

The implication is quite obvious. The vision and strategies can be well formulated, but when the 3P management and benefits realisation are not in place, then all the work is done in vain. The implication is that the ship will go in circles and not reach the final destination. Or, even worse, the ship might sink somewhere along the line.

Two other concepts that need to be considered are that of project management sustainability and maturity. Sustainability should be part and parcel of the comprehensive 3PM framework, not just an addition to it. The ultimate goal is that organisational leaders ensure that the principles of sustainability are incorporated within the portfolio at large. As with governance, sustainability should be a culture within the organisation that is entrenched in the day-to-day management of a project.

A goal that organisational leaders strive toward is that of excellence. This is also applicable within the 3P environment in which excellence can be related to the achievement of project management maturity. Project management maturity is achieved when organisational leaders have all the enablers in place to aid 3P managers to achieve maturity.

When all the concepts of the framework are in place and performing optimally, then the loop can be closed where value is achieved through project management. The question that is raised is why organisational leaders should support project management, and the answer lies in the value that is created. Project management value is created when projects ensure organisational success.

12.3 Chief Project Officer

Organisational leaders should seriously start thinking of introducing the role of a Chief Project Officer (CPO). This individual can oversee the entire comprehensive 3PM framework. The CPO will be part of the C-suite and report on the progress of strategy realisation through the implementation of the portfolio, programmes, and projects. The role of the CPO will fluctuate between accountability and strategy execution.

The CPO will be ultimately responsible for the entire value chain, as depicted in Figure 12-1. This implies that as the guardian of the value chain, the CPO is accountable for the portfolio and how programmes and projects are selected and implemented. This accountability will stop the debate as to who is ultimately responsible and accountable for the realisation of benefits. The CPO will also take responsibility for the improvement of organisational project management maturity (PMM).

Although every person within the C-suite is accountable for strategy implementation, the CPO can take a leading role in this regard. Since projects are perceived as the vehicles to implement strategy, the CPO can be the owner and report on the progress with regard to realising the organisational strategies and vision on a frequent basis.

At the end of the day, the CPO will be the one-stop person for all project management–related issues, concerns, and victories.

12.4 Conclusion

This book took us on a journey to discover why project management is important and why organisational leaders should embark on and support project management. It is evident from the framework that project management itself is complex and that it becomes even more complex when it is placed within the context of an organisation.

It is evident that when organisational leaders want to deliver the vision and strategies of the organisation, the only vehicle is that of project management. Project management *per se* cannot be seen in isolation but needs to be seen in relation to portfolio management and the internal workings of the project itself. Organisational leaders need to be aware of this complexity, and the 3PM framework tries to simplify this complex integration. Recognising the complexity of project management empowers organisational leaders to make the right decisions as well as choices for the organisation.

This framework also makes organisational leaders aware of the various individual concepts that contribute to project management and that a holistic approach is required to manage projects, reap benefits from projects, and ensure the success of the organisation.

Some final advice is that familiarity breeds comfort. When organisational leaders are familiar with project management and eveything that it entails, then the sense of "I don't understand, I'm nervous, I'm out of control" disappears. This comfort allows them to embrace project management as a vehicle for organisational success.

List of Acronyms

APMBOK	*APM Body of Knowledge*
BDN	Benefits Dependency Network
BRM	Benefits Realisation Management
BRP	Benefits Realisation Plan
CPO	Chief Portfolio Officer
CSR	Corporate Social Responsibility
GAPPS	Global Alliance for Project Performance Standards
GRI	Global Reporting Initiative
HSE	Health, safety, and environment
ICB®	International Competence Baseline
ICT	Information and Communications Technology
ISO®	International Organization for Standardization
IT	Information Technology
ODE	*Oxford Dictionary of English*
OPM3®	*The Organizational Project Management Maturity Model*
P3	Project, programme, and portfolio
PfM	Portfolio Management
PgM	Programme Management
PM	Project Management
PMCDF	*Project Management Competency Development Framework*
PMI®	Project Management Institute

PMO	Project Management Office
SD	Sustainability Development
SIN	Sustainability Impact Network
V2P	Vision-to-Project Framework
PMMM	Project Management Maturity Models
PRINCE2®	PRojects IN Controlled Environments
WBS	Work breakdown structure

Index

C

G

H

I

J

K

(*continued on following page*)

P

Q

R

S

(*continued on following page*)

T